CREATING ORIGINAL PROGRAMMING FOR CABLE TV

Edited by Wm. Drew Shaffer
and Richard Wheelwright
for the *National Federation of
Local Cable Programmers*

Executive Editor
Mary Louise Hollowell

A15043 397063

Communications Press, Inc.
Washington, D.C.

Grateful acknowledgment is made for permission to print the following: "Basic Legal Planning for the Producer of Programming, or I Wish I Had Thought of That Before We Started Production," © 1983 by Ernest T. Sanchez; the forms provided by U.S. Cable of Northern Indiana in "Cable Television in Advertising," by Margie Nicholson; "Advertising and Marketing on Cable Television: Whither the Public Interest?" by Dee Pridgen and Eric Engel, from *Catholic University Law Review, Vol. 31, Winter 1982, No. 2,* © 1982 by The Catholic University of America Press, Inc.; and the Glossary, compiled by Mary Louise Hollowell, © 1980, 1983 by Communications Press, Inc.

ISBN 0-89461-036-8

Copyright © 1983 by National Federation of Local Cable Programmers

Published by Communications Press, Inc.

All rights reserved. Not to be reproduced in any form whatsoever without written permission from Communications Press, Inc., 1346 Connecticut Avenue, N.W., Washington, D.C. 20036.

Printed in the United States of America

Cover design by Linda McKnight

Second printing December 1984

FOREWORD

Since 1976 the National Federation of Local Cable Programmers has provided a mechanism by which individuals, community organizations and institutions, and local governments could share their successes and failures in dealing with community programming on cable systems.

The National Federation of Local Cable Programmers (NFLCP) was organized to foster citizen participation in community television programming. The goals of the NFLCP are to discover and assist users of local channels, to facilitate the exchange of information between people throughout the country who are concerned with community responsive programming, and to spread innovative programming ideas among community access centers. NFLCP works to achieve these goals through regional and national workshop and training programs. In addition, it publishes and/or distributes a growing number of publication and videotapes. *Creating Original Programming for Cable Television* is our newest effort.

We are grateful to Drew Shaffer for his patience and dedication in the editing of this book; to Mary Louise Hollowell and Communications Press for their support and belief in this project; and to all the authors for their insights and experiences.

Sue Miller Buske

Executive Director
National Federation of
Local Cable Programmers

CONTENTS

Foreword iii
Sue Miller Buske

1. Introduction—Overview 1
   Drew Shaffer

Part I. CREATING ORIGINAL PROGRAMMING FOR CABLE TV: MAKING IT WORK FOR YOU

2. Programming Sources 11
   Susan Wallace

3. Producing Original Programming for Cable 21
   Ann McIntosh and Steve Feldman

4. Independents and Access: The Status,
   The Predicament, and The Promise 34
   Fred Johnson and Drew Shaffer

5. Basic Legal Planning for the Producer of
   Programming, or I Wish I Had Thought of That
   Before We Started Production 44
   Ernest T. Sanchez, Esq.

6. Making the Connection: Distributing
   Original Programming 53
   Kitty Morgan

7. Cable Television Advertising 65
   Margie Nicholson

8. How the Cable System in Iowa City Works—
   Toward An Interactive Future 85
   Richard Wheelwright and Drew Shaffer

Part II. CREATING ORIGINAL PROGRAMMING FOR CABLE TV: AVOIDING PUBLIC INJURY AND COSTLY REGULATION

9. Advertising and Marketing on Cable Television:
   Whither the Public Interest? 95
   Dee Pridgen, with Eric Engel

Glossary 143

Wm. Drew Shaffer

1. Introduction—Overview

In 1982 cable television represents one of the fastest growing and most dynamic industries in the United States. In a frenzy of activity, and with substantial investment, it is physically connecting communities and cities with each other, in a web of communications conduit. Feeding into this conduit are many other dynamic forms of media and transmission systems resulting in a national communications network. This network is creating a multiplicity of cable channels, or conduit outlets, and a need for information and programming to fill those channels— on local, regional, and national levels.

Many forces are at work, influencing and shaping the form, function, and content of these channels and programming. These forces are the "players" that ultimately determine what we see on our television sets via cable.

Wm. Drew Shaffer is the Broadband Telecommunications Specialist for the City of Iowa City, Iowa. Mr. Shaffer is a producer of educational, fictional, documentary and commercial work for such institutions as NEH, NCAT, and the Iowa Governor's Safety Council. He has also been employed as a video and cable consultant since 1972 and has taught at the University of Iowa.

Copyright 1983 by National Federation of Local Cable Programmers (NFLCP).

THE PLAYERS IN CABLE

The players of cable include the conduit providers; those who own the forms of distribution, and who are responsible for insuring the necessary hardware is in place so the signals can be sent. The players also include all those involved in influencing and molding the function or uses of the cable channels, such as the cable companies, legislators, lobbyists, regulators, and courts. Finally are the content or programming suppliers—those persons involved in creating, producing, facilitating the producing of, and distributing programming. Simply put, the makers and owners of the "wire"; the controllers of the wire; and the producers of the programming on the wire.

The Cable Providers

Many corporations and individuals work behind the scenes to enable us to receive cable. In order for any cable system to exist, manufacturers have had to produce all the necessary equipment, such as the strands, cables, modulators, splitters, taps, amplifiers, and converter boxes. Before the cable system can be activated, subcontractors have to be hired to map and construct the system. On the local level the providers of cable are the cable company; the manager, the system engineer, the technicians, and the installers. On the local, regional and national levels the owners of satellite, broadcast, computer, microwave, and telephone systems provide the hardware (and in some cases the software) to get the signals to the cable operator. In 1982, these are the necessary components which work together to bring cable and its services to your home.

Even five years ago the number, size, and roles of these components was small and/or easily identifiable. As the business of and money in cable grew, so did the number and size of cable providers. This tremendous growth has caused shifts in roles as well as power. Cable companies, once only wire providers, are now program providers as well. Many cable companies have grown from small, one-town outfits to multiple system operators (MSOs) some of which are multinational conglomerates. Several own over one hundred cable systems, with each system capable of millions of dollars of revenue per year, and simultaneously own radio, television, and newspaper chains. The networks of ABC, NBC, and CBS, once the proprietors and gatekeepers of television, have been relegated to a choice among dozens, rather than one of three. WGN, WTBS, and WOR, once independent stations reaching their respective state's residents, have been catapulted to "superstation" status now that they're carried nationwide by satellite.

Today these shifts in roles and power are still dynamic and changing. More interconnection possibilities are being generated betweeen cable and non-cable media and transmission systems, leading to more channel outlets and more programming possibilities. These cable providers have played an integral role in the creation of what cable telecommunications represents today and what it will represent tomorrow. Many have also played a role in controlling cable.

The Cable Controllers

The cable controllers include municipal, state and federal legislators and regulators, the courts; and lobbying groups. Some of those involved in cable-related lobbying include: the cable companies, in the form of the National Cable Television Association (NCTA); and public or public interest groups such as the National League of Cities (NLC); National Citizens Committee for Broadcasting (NCCB); National Federation of Local Cable Programmers (NFLCP); and Association of Independent Video and Film Makers (AIVF). A number of broadcast lobbyists, such as the National Association of Broadcasters (NAB) and the telephone companies have also lobbied on cable-related issues. Each has played a role in determining what programs will be seen where, and when, on cable.

Very little has been legislated in the area of communications since the 1934 Communications Act. This Act provided many of the basic principles that have regulated all communications media, even though many of the media forms discussed in this book were not invented until long after its enactment. The intentions of the Act were clear, and among them were: to enable the creation and operation of a national communications system at the lowest possible cost; to emphasize localism by encouraging communications on the local level; to make clear that the airwaves belonged to the public; and that broadcasting was a privilege that could be revoked. These are very noble ideas that have only in inconsistent degrees been passed on to succeeding generations of media regulation, and have always been a matter of debate, primarily between the communications industries and public interest groups.

Federal legislation and regulation remained, for the most part, inert between 1934 and 1982 on the matter of cable, except for the Federal Communications Commission (FCC) in the late 1960s and early 1970s. Here the FCC's activity primarily involved the mandating of public access to cable systems and the relaxation of certain cable restrictions, which has allowed the industry to grow. By the late 1970s thirteen states and hundreds of cities had begun to legislate and regulate cable. Without federal guidance and their own expertise in cable, much of this state and local legislation and regulation has remained under debate or attack, largely because of claims of inconsistency and/or unfairness (from both the cities and cable companies). For the last three years there have been industry-sponsored Senate bills attempting to remove municipal and state cable regulation (in 1980 it was S.B. 2827; in 1981, S.B. 898; in 1982, S.B. 2172) primarily with arguments of first amendment rights of cable companies. The cable companies want total discretion to choose whatever is put on their cable systems. The bills' opponents argue cable should be treated as a common carrier, public utility, or a hybrid entity. These bills have not passed—yet.

The courts have entered the battlefield as well, most notably in 1972 and 1978 in what have become known as the Midwest Video Corp. vs. FCC I and II cases. Both cases involved public access to cable system mandates. The 1972 case upheld FCC access requirements, and the 1978 case struck down the access requirements on the grounds the FCC had not been granted the authority to make such requirements (the only major bill in effect being the 1934 Communications Act). Most recently the courts have refereed city and cable company disputes in Community Communications Corp. (owned by Tele-Communications, Inc.) vs. City of Boulder in 1980 and in the suing of the City of Tucson, Arizona, by Catalina CableVision in 1982. Both situations involve cable-initiated antitrust cases against these cities, alledging undue restraint of competition.

These issues, as well as many others, remain unresolved. Their resolution will determine who ultimately controls what is put on the cable systems, which, of course, affect the producers and providers of cable programming.

The balance of this book deals, for the most part, with the cable programming producers and suppliers. The preceding sections on the cable providers and the cable controllers are not meant to provide comprehensive coverage of these entities, events, and influences, but rather to illustrate that creation of original programming for cable television does not exist in a vacuum, and cannot be dealt with in one. In fact, what is and can be produced and distributed is influenced to a great degree by these entities. It is within this context that creating and distributing original programming for cable television can and should be broached. (Please see the Resources section at the end of this chapter for references to comprehensive coverage of these areas.)

The Cable Programming Producers and Suppliers

The programming producers and suppliers are all the writers, actors, technicians, directors, producers (independent, community and commercial), entrepreneurs, innovators, advertisers, businesses, production houses, access centers, cable companies, cities, libraries, hospitals, schools, colleges, citizens, community organizations, distributors, networks, associations and lawyers that create, produce, and facilitate the production and distribution of programming on local, regional, and national levels.

There are, as you may have noted, several new additions to what previously may have been considered a list of programmers or producers (such as cities, libraries, and schools). This is indicative of current discussion involving the new possibilities, opportunities, and arrangements in creating original programming for cable television. There are many newcomers because of what cable television now offers in the way of more channels; and because of this cable can now begin to accommodate the diverse opinions and needs that all these players represent. On the local level cities, libraries, schools, churches, hospitals, citizens, independent producers, community organizations, cable companies, businesses, and many others are beginning to communicate visually via cable. This is a new experience for most of these players; an uncharted area with precious few blueprints to follow. Commercial opportunities on the local level are equally viable, yet just as uncharted. The demands, needs, and applications are in the process of being determined. Experimentation and variation in approaches to cable are the order of the day.

An independent producer may attempt to produce commercials for his/her own local show on the local origination channel. A citizen may produce an informational program about the organization he/she belongs to for an access channel. The cable company may create local revenue by setting up an advertising program on its channel. Schools and hospitals may interconnect with the library and utilize the institutional network to search catalog listings or make data transactions. Cities have started to provide public service information on government access channels. Innovators are approaching cable and finding outlets to experiment with the format and style of television such as changing its function to a more participational or interactive medium. The list goes on and on.

Regional interconnection of cable systems and cities, happening in many areas such as Minneapolis and Cincinnati, will allow for an extension of this expression and communication, and at the same time will provide communications of a different nature and wider scope.

National program services and outlets for programming to cable systems have multiplied exponentially with the advent of satellite delivery. The nature of this programming again, if anything, represents diversity. More resources and opportunities on the local level (such as

6 / Creating Original Programming for Cable TV

access centers and lower-cost production equipment) allow for the possibility of producing programming for national audiences. Various combinations of independent producers, directors, writers, actors, and entrepreneurs scramble to produce programs for satellite, pay, or basic services. Marketing, distributing, and copyrighting considerations become very important factors in success or failure, profit or loss. Once again, the avenues are unclear and the choices are many.

Cable operators try to determine what satellite services to carry and how many cable channels to reserve. Brokers buy and sell satellite time and make a living. Whole satellite transponders are bought, auctioned, and reserved at phenomenal fees.

These players are working in a variety of arrangements (non-profit, cooperative, and commercial), on a variety of levels (locally, regionally, and nationally), from a variety of perspectives, intentions, and motivations for different economic, political, philosophical, artistic, spiritual, and social reasons. The best way to exemplify this statement and the current status of those involved in programming, is that it represents a state of transition. Because of dramatic changes—like those of cable availability to virtually all cities; cable channel growth; satellite delivery; and several others; things that couldn't have occurred before, are occurring now. What hasn't happened yet can or soon will. This transitional phase means changes are occurring, and the changes mean opportunities for those who seek them. It represents a period when all the ground rules are not yet thought of, made, or agreed upon; when power and control shifts occur daily. In fact, the best and most challenging conditions, for exploration by the players in cable, are here now.

THE ROLE OF THIS BOOK

Each chapter of this book offers insights into how this transition period and its uncharted areas are being or can be navigated. Each chapter stands on its own. That is, a single chapter may fulfill one reader's needs, while another reader may benefit from all the chapters. Someone who knows little about cable in general can learn much while someone who is an expert in one cable area (for example, program production) can expand his/her knowledge into a new area (for instance, advertising).

This book is written by players in cable; people writing about what they do in cable-related jobs. Their perspectives and styles have largely been left intact, because this approach best illustrates what cable represents—diversity.

Introduction—Overview / 7

RESOURCES

Cable, Computers, and Interactive Services and Experiments

1. Hollowell, Mary Louise (ed.), *The Cable/Broadband Communications Book Volume 2, 1980-1981,* Communications Press, Inc., 1346 Connecticut Avenue, N.W., Washington, D.C. 20036, 1981.
2. Jesuale, Nancy and Smith, Ralph Lee (eds.), *The Community Medium,* The Cable Television Information Center, 1800 North Kent Street, Suite 1007, Arlington, Virginia 22209, 1982.

Cable and New Technologies

1. Alnes, Steve, *Telecommunications: A Picture of Change,* Upper Midwest Council, 250 Marquette Avenue, Minneapolis, Minnesota 55480, 1981.
2. *Catalog of Publications and Events on Television, Private Non-Broadcast Video, Home Video and Cable, Videotext,* Knowledge Industries Publications, Inc., 701 Westchester Avenue, White Plains, New York 10604, 1982.
3. *Telescan,* American Association for Higher Education, One Dupont Circle, Suite 600, Washington, D.C. 20036.

Cable Company Management, Marketing and Advertising

1. *CableNews,* Phillips Publishing, Inc., Suite 1200N, 7315 Wisconsin Avenue, Bethesda, Maryland 20814.

Cable and Programming Issues

1. *Channels,* Media Commentary Council, Inc., Channels of Communication, Box 2001, Mahopac, New York 10541.

Cable Industry Trade Publication

1. *CableVision,* Titsch Communications, Inc., 2500 Curtis Street, Denver, Colorado 80205.

Cable Regulation on City, State, and Federal Levels/Ownership Issues

1. Hollowell, Mary Louise (ed.), *The Cable/Broadband Communications Book Volume 2, 1980-1981,* Communications Press, Inc., 1346 Connecticut Avenue, N.W., Washington, D.C. 20036.
2. Jesuale, Nancy, Neustadt, Richard M., and Miller, Nicholas, *A Guide For Local Policy,* The Cable Television Information Center, 1800 North Kent Street, Suite 1007, Arlington, Virginia 22209.

8 / Creating Original Programming for Cable TV

City Uses of Cable

1. *Management Information Service Report, Volume 14, Number 6,* International City Management Association, 1120 B Street, N.W., Washington, D.C. 20005, June, 1982.
2. NATOA (National Association of Telecommunications Officers and Advisors), National League of Cities, 1301 Pennsylvania Avenue, N.W., Washington, D.C. 20004.
3. Orton, Barry (ed), *Cable Television and the Cities: Local Regulation and Municipal Uses,* University of Wisconsin-Extension, Madison, Wisconsin 53706, 1982.

Distributors, Program Listings, and Wholesalers

1. *The Sixth International Video Exchange Directory,* Satellite Video Exchange Society, 261 Powell Street, Vancouver, British Colum- Canada V6A 1G3, 1979.
2. *Video Source Book,* The National Video Clearinghouse, Inc., 100 Lafayette Drive, Syosset, New York 11791, 1980.

Guide to Incorporating

Nicholas, Ted, *How to Form Your Own Corporation Without A Lawyer For Under $50,* Enterprise Publishing Company, Inc., 1300 Market Street, Wilmington, Delaware 19801, 1981.

Independent Producer Opportunities

Gadney, Alan, *Gadney's Guides to 1800 International Contests, Festivals and Grants,* Festival Publications, P.O. Box 10180, Glendale, California 91209, 1980.

Public Access and Community Programming

Community Television Review (CTR), University Community Video, Inc., 425 Ontario, S.E., Minneapolis, Minnesota 55414.

Public Interest Communications Lobbying and Legislation

Access, National Citizens Committee for Broadcasting, P.O. Box 12038, Washington, D.C. 10005.

Technical Aspects of Cable

Cunningham, John E., *Cable Television,* Howard W. Sams and Company, Inc., 4300 West 62nd Street, Indianapolis, Indiana 46268, 1980.

Part I.
Creating Original Programming for Cable TV:
Making It Work for You

Susan Wallace

2. Programming Sources

The cable operator today is faced with many programming related dilemmas. While many of the cable systems built prior to 1978 are twelve channel capacity operations, cable systems under construction or being rebuilt may have as many as 108 channels. The average cable system built today will have 36 channels. Not only have the number of cable channels multiplied, but so have the programming sources for these channels.

The innovation of satellite programming services has dramatically increased what is available to the cable operator in the way of programming. In fact, it is the availability of satellite channels and programming that has given the cable companies the capability of offering and filling more than a twelve channel system. The growth of these satellite services has been phenomenal since 1979, and continues at a tremendous rate. Every time a cable operator picks up a trade publication, yet another satellite service is available. These satellite services vary greatly in content, format, and cost.

In addition to the satellite services available to cable operators, a number of channels have been set aside by local cable ordinance or

---

Susan Wallace is the Director of Advertising Sales and Community Relations for Metrovision, Inc., of Atlanta, Georgia. The views expressed herein are those of the author.
Copyright 1983 by National Federation of Local Cable Programmers (NFLCP).

franchise that provide additional programming for the local community. One category of such channels is referred to as access or community channels. Programming on these non-profit channels is produced by community individuals, organizations, institutions, the cable company and independent producers. Sometimes these channels will be a mixture of locally produced material with locally relevant programming produced outside the community or with satellite programming. Another category of local channels is called leased channels. A leased channel is one that has been reserved for local commercial or entrepreneurial purposes. The cable company may also have a local origination channel. This is the cable company's potentially commercial channel which can be used to exhibit it's own productions or other programming as it sees fit.

Also available to cable operators is a variety of automated news and information services. Such services are usually brought to the cable operator via telephone lines and make up from ten to twenty-five percent of the total cable television menu.

Over-the-air broadcast stations in the immediate vicinity of the cable operator, or "must carries", make up the balance of programming sources for the cable operator.

Let's take a closer look at each of these individual sources of programming.

SATELLITE SERVICES

Non-Premium or Basic Services

As of August 1, 1982, no fewer than thirty-one non-premium or non-pay and basic services are available to the cable operator via satellite. Non-premium, non-pay, and basic services refer to all the programming offered by the cable operator on the basic tier of cable service. Such services vary considerably in content, format, and cost to the operator.

The content of non-premium satellite programming spans a considerable range of subjects. Some services include: an all women's channel called The Women's Channel; an all black channel called BET; several religious channels such as CBN or PTL; sports channels, for instance, ESPN and USA Network; Nickelodian, a children's channel; public affairs channels such as ACSN or C-SPAN; CNN or SNC news channels; a music channel, MTV; ARTS, an example of an arts channel; and foreign language channels such as SIN. "Superstations," or independent broadcast stations are carried by satellite as well, including WTBS in Atlanta, WGN in Chicago, and WOR in New York. Finally, some systems may choose basic audio services such as Lifestyle and WFMT.

Some of these channels provide programming twenty-four hours a day, as with CNN or MTV, while others, for instance, BET and MSN, program part of the day. Those channels that program in limited time blocks are often combined with other programming or channels to maximize channel usage. Sometimes they are combined with a local access channel or local originating channel and sometimes with another satellite-fed channel. For example, MSN may be combined with the Episcopal TV Network.

The cost of this satellite programming varies as much as the content and format. A few services, such as ACSN or CBN are free. Most services range in cost from two cents to seventeen cents per cable subscriber. The cable operator pays these fees and incorporates the cost into the basic cable service charge.

Most of the basic service satellite offerings are advertiser supported. Some services, such as ESPN, have as many as 173 advertisers, over half of which are new advertisers in the last year. Forty-two of these advertisers are new to network advertising. Such figures and ratios are not unusual for the satellite-offered services. Considerable interest in this form and delivery of programming has been shown by the Fortune 500 companies as well as by smaller businesses. While such advertising revenue fuels the current satellite services and promotes the birth of new services, it opens very attractive opportunities to the independent producer (someone has to produce the programming and commercials for all these new services).

The average cable system will carry one or two independent superstations, at least one sports package, and one religious channel. The cable operator may then combine some part-time channels to fill unused channel space, while leaving room for at least one pay channel service.

Pay or Premium Services

In addition to the thirty non-premium services eleven pay or premium services are available by satellite. These pay services are offered as a second (or more) tier of programming with the additional charge added to the basic service fee. Sometimes this charge is more than the basic service fee; sometimes rather than a flat monthly rate, the cost is determined on a per viewing basis. A second characteristic of most pay channels is that they are not advertiser supported and therefore no advertisements are run.

Most pay channels, for instance The Movie Channel, HBO and Home Theatre Network, provide round-the-clock fare. One pay channel, Bravo!, is characterized as a cultural service offering concerts and ballet, and another, Escapade, offers adult-oriented programming. The quality of movies presented varies from first run, made for cable distribution, to second or third run movies to the classic films of the 1940s and 1950s.

The average cable system offers between one and four pay services. The new builds with as many as 108 channels offer almost as many pay services as are available. Pay services continue to be added by the cable operator because they are a cable operator's bread and butter. The revenue from these services makes cable TV the lucrative investment opportunity it is. Besides the promising pay programming growth projections, pay cable subscribers are expected to multiply six times from 1980 to 1990, increasing to over 43 million.

Most of the pay channels are owned by an MSO (Multiple System Operator). For example, American Television and Communications Corporation (ATC), the second largest cable company in the country, owning over 130 cable systems, is owned by Time, Inc. Time, Inc. also owns HBO and Cinemax, two satellite pay channel services. It is not surprising that in many systems owned by MSOs, the MSO's own pay channels are the only ones offered.

Future Satellite Services

Most of the satellite pay channel services, as well as the non-premium services, are carried on three satellites: Satcom IIIR, Satcom IV, and Westar IV. As many as seventeen more satellites will be in orbit by 1985. Some of these satellites will have as many as twenty-four transponders, with each transponder capable of carrying one programming service such as CNN or many audio or data services. These seventeen satellites will be financed by eight companies, including RCA Americom, Western Union, AT&T, GTE, Hughes Communications, Satellite Business Systems, Southern Pacific Communications, and Space Communications Co.

Satellite channel or programming service growth is projected to continue its skyward trend, along with the satellites themselves. Forty-five more satellite services are proposed for 1983 and 1984. Many of these proposed channels are basic services, while at least six are pay channels. Much of the content of these channels is exemplified in their names, and spearheads a dramatic change in nationally offered programming for specialized audiences, for example: the Cable Health Network; the Cable Newspaper Corp.; The Job Channel; Magicable; Nostalgia Network; and Shopping by Satellite.

SOURCES OF PROGRAMMING FOR THE LOCAL CABLE CHANNELS

Access or Community Channels

The access or community channels may be programmed in a variety of ways, depending on several variables. If there is strong city, company, and community support for these channels, the workshops, equipment, and staff necessary to assist the community producers in producing their programming can result in strong, successful access channels. In many such instances local access channels cablecast only locally produced material a significant portion of the day. One of the earliest uses of community programming was to bring Mass to the homebound. In 1982 programming is produced by organizations, individuals, church groups, schools, self-help groups, libraries, and government agencies from all levels, ages, occupations, and socio-economic status. In some cities franchised since 1979, such as New Orleans, the cable company offers grants for the production of local programs. Anyone can apply for and get these grants to produce programming of local relevancy. In addition, some access channels permit some form of underwriting of programs. This allows community producers to get sponsorship and thus reimbursement for their work.

The number of these channels varies according to cable system capacity, size of community, ordinance requirements, or franchise-proposed system design. One to four access channels is fairly common. If there is only one access channel, participants share a composite channel. As channel capacity expands, there may be separate channels for the public, government, schools, and library. These channels may be operated and administered by a non-profit organization created just for that purpose, a governing body or consortium (i.e. the school channel may be run by a consortium of educational institutional representatives). In other cases, especially in smaller communities where only one access channel exists, the cable company operates the access channel. Minimally, a period of two to three years is required to develop such channels to the point of being programmed a significant portion of the day with local programming, once the cable system is constructed.

In the early stages of access channel development, to build viewer support (or if the makeup of the community does not lead to locally produced material), the access channel may take the form of a theme channel. Theme channels mix available local programming with a variety of other programming sources.

Programming of local interest which is produced on a regional, state or national level may be combined with locally produced programming on the access channels. This is done to fill channel space with worthwhile programming and build audience viewership. A variety of organizations such as Ducks Unlimited offer travelogues. Health programming from medical schools, Public Health Departments, the American Red Cross, and the American Cancer Society is available. Other program sources include state universities, community colleges, state boards of education, state capitals (some states do programming from the state capital like the Illinois Press Show, congressional representatives, or the United States government. The U.S. government produces tapes on everything from how to avoid botulism in canning to why we have the best bomber planes in the world. The U.S. Information Agency is one place to locate such programming. Most such programming is free or available at a low cost, and is usually copyright cleared for cablecasting. Programs that require a fee or are not copyright cleared are usually not played on access channels. As the cable education process expands, more public service programs and producers will be aware of and allow for the clearances needed for cablecasting. The decisions about which of such programs to use are left to the administering agency. As the number of community programs being produced increases, usually the amount of non-locally produced programs being cablecast decreases.

Theme channels are sometimes made by using well known satellite overflow programs that can be placed on access channels to supplement the number of hours of programming available and to increase the awareness in the community of certain national events. An example of this type of programming would be the SALT II talks from Vienna that PBS carried from the Westar satellite. Because no PBS stations wanted to run these shows for their scheduled fifteen hour length, they were made available to cable operators for cablecasting in their entirety over community or local origination channels. Another example of satellite overflow programs comes from NASA. In 1982 NASA made available the Jupiter Fly By, the Saturn Fly By, and the Space Shuttle Landing to operators for transmission on community or local origination channels.

Many cable operators support access or community channel development because of ordinance or franchise requirements and/or because they believe such channels and programming strengthens their cable service. Studies done in Iowa City, Iowa, and Kettering, Ohio, indicate that at least five to eight percent of the cable TV audience is attracted by just such programming. The cable operator would like to believe community programming is a service bought by subscribers for some of the same reasons they buy ESPN or any other cable service. Other cable operators direct much attention and energy to a local origination channel, again reflecting the thinking that by doing so, the strength of the service is enhanced, and because local revenue can be created.

Local Origination Channels

Depending on the city, community, and/or cable company, a potentially commercial local origination (L.O.) channel operated by the cable company may develop. If the cable company is required to assist in access channel development, this may impede the L.O. channel's development for several years. Only so much staff, equipment, and energy can be economically justified for such purposes by the company in a given community. In any event, L.O. channels have or will be developed in most of the larger cable systems by 1985.

Because L.O. channels can be commercial channels, the potential for commercially-sponsored local productions done by the company, independent producers, or others is becoming a real possibility. The L.O. channel allows the smaller businesses an opportunity to advertise on TV; an opportunity previously lost to the high cost of broadcast TV advertising. In contrast the cost of cable TV advertising is comparable to radio rates.

Frequently in early stages of development the L.O. channels will exhibit theme channel characteristics. Because revenue can be created easier than access channels, more money can be spent on programming that is cablecast. Consequently programming that is of low cost may be purchased or a negotiated trade-out may be arranged. A tradeout is a service-for-service exchange. For instance, the cable company may do a commercial for a lumber company in exchange for wood to build a new set design. Sometimes a satellite-fed service with local (commercial insert) availabilities may be cablecast on the L.O. channel and local commercials will be produced to create more revenue. Sometimes this is done by the cable company and sometimes by a local entrepreneur. Such scenarios are potentially more lucrative and possible as a result of the satellite-fed programming. It is still impractical for most cable operators to produce on their own a comparative amount of programming to fill their L.O. channel.

Leased Channels

In some twelve channel systems and in nearly all cable systems with thirty-five or more channels at least one leased channel (sometimes called a leased access channel) is set aside. Usually this channel is set aside for unspecified or future development of commercial or entrepreneurial applications. Such channels could be used for fire and burglar protection systems, data transmission or transactional purposes for banks or other institutions. In some places the leased channel is rented, in part or all, by entrepreneurs. In one such instance an entrepreneur rented part of a leased channel for two hours a day every day of the week, produced music and talk programs, found the necessary sponsors, and produced the sponsor's commercials to be inserted in the programs.

The costs for the use of the leased channel vary and are subject to negotiation. In some cases a flat fee is charged. In others a percentage of the gross or net profit is charged, or a trade-out arranged.

Automated Channels

An additional source of programming for the cable operator is the automated channel. Automated channels are an economical supplement to the cable system's offerings, particularly in thirty-five or more channel systems where so many channels have to be filled. These channels convey data, graphic or slow-scan images with topics like state and national news, stock exchange, weather, and time information.

All of these services are usually delivered to the cable operator via telephone lines. Although some of the services themselves are free, all have associated costs. For instance, while many weather radar services are free (supplied by the National Oceanographic and Aeronautics Administration) it may cost the operator $400 dollars per month to bring the signal to the locality via phone lines. Data news services may be supplied by Associated Press (AP) at another $400 per month. Compared to other programming costs, these prices are nominal, yet they are an added expense. If many data channels are supplied the cumulative price becomes significant.

"Must-Carry" Broadcast Stations

Must-carry stations complete the programming sources available to a cable operator. Must-carries are broadcast stations or channels that are designated by the Federal Communications Commission (FCC) as significantly viewed in that area and therefore must be carried by the cable company. The number of must carries in any given cable system depends on the number of broadcast stations viewable over the air in that area. Such a factor may play heavily in programming choices left to the cable operator and consequently the audience. In a city such as New York, where there are six broadcast TV stations, both network affiliates and independents, the number of must-carries may become significant. This is also true in the Washington, D.C./Baltimore area where municipalities may require the cable operator to carry both the Washington, D.C. and Baltimore stations. Six FCC designated must-carries in a twelve channel capacity system severely limits what other programming may be put on that system. From the point of view of the community producer of access programming, getting programming on such a system may be difficult, and may become increasingly difficult as channel competition becomes fierce.

SUMMARY

There are an enormous amount of programming sources available to the cable operator, and many factors involved in determining which programming sources will end up on the cable system. Satellite-delivered programming brings a host of new viewing alternatives to the cable operator, and in some cases creates new revenue possibilities. These programming sources and the number of satellites delivering them are increasing rapidly. While the older and smaller twelve channel systems may not have channel space for this wealth of material, the new franchises and the rebuilds offering thirty-five or more channels will have a plethora of choices.

Local cable channels also bring a variety of programming sources to the cable system. Access channels deliver locally produced material or may combine local programming with other shows from sources outside the community. L.O. channels offer local commercially-produced program opportunities. Leased channels open new possibilities for the company, entrepreneur, and independent producer to deliver new and experimental programming and services. Automated channels and must-carry broadcast stations round out the programming sources available to the cable operator.

Each community cabled has its unique blend of visual offerings, some of which are dictated by the community and the FCC. Other programming is the choice of the cable operator, or of the access channel administrators, or is illustrative of community or entrepreneurial imagination. All these programming sources are being combined in cable systems across the country, offering tremendous viewing, producing and revenue-creating opportunities for the cable audience, the producers of programming and services, and the cable operator. If industry projections are accurate, there is much more on the horizon.

RESOURCES

Cable File, Titsch Communications, Inc., 2500 Curtis Street, Denver, Colorado 80205, 1982.

The Cable Television Factbook (Services Volume), Television Digest, Inc., 1836 Jefferson Place, N.W., Washington, D.C. 20036, 1981.

CableVision, Titsch Communications, Inc., 2500 Curtis Street, Denver, Colorado 80205.

Ann McIntosh and Steve Feldman

3. Producing Original Programming for Cable

A variety of developments, including the growth of cable systems and the penetration of those systems, the multiplicity of channels on cable, regional interconnection of cable systems, and satellite delivery of cable programming, have brought new variables into the "formula" of programming, programming possibilities and opportunities, and distributing, for everyone involved in these endeavors. It is to every producer's benefit to consider the thrust of these variables, which can be characterized by the following questions: Do more channels mean more programming will be made and bought? If so, from whom, by whom, and in what kinds of arrangements? Do more channels mean more of the same kind of programming we've seen on broadcast television for thirty years, utilizing the formula principles, or can different, or better,

---

Ann McIntosh is currently the Access Implementation Planning Consultant to the Boston Access and Programming Foundation. She is also Co-Director of Lobster Associates, Inc., a regional cable programming and distribution company. Ms. McIntosh has worked as Program Director for Times-Mirror in Brookline, Massachusetts.

Steve Feldman is Co-Director of Lobster Associates, Inc. He has previously worked as Program Director for Continental Cablevision in Newton, Massachusetts; as producer/director for South Carolina Educational Television; and as a video artist at WGBH.

Copyright 1983 by National Federation of Local Cable Programmers (NFLCP).

programming be made, sold, and distributed? How can producers create strategies that will enable the funding, producing, and distributing of their work through these new outlets?

There are three definite, and very different, markets for producing programming for cable television. There are the local, regional, and national markets. Each has its own unique set of characteristics and possibilities.

THE LOCAL CABLE MARKET

The local cable channels, such as access, local origination, and leased access channels, offer the most accessible route for producers to begin to create, experiment with, and distribute their work.

Many of the access channels are operated by access centers created through cable company franchises with cities. Those centers created since 1979 offer a reasonably sophisticated kind of hardware, commonly referred to as "industrial," including: three-tube, prism optic cameras and three-quarter inch videotape recorders; half-inch VHS or Betamax equipment; studios with video switchers and audio mixers more than adequate to produce complex dramatic or musical tapes in aesthetically innovative ways; and editing systems with time coding and time base correction. Much of the equipment in these centers is capable of producing programming for virtually any market. Chapter Four, "Independents and Access", concentrates on how to work with these centers, and the non-profit and cooperative arrangements that producers can make with access centers for programming and distribution purposes. This section concentrates on the variables of unique program development, and distribution and feedback advantages of local access cable channels, the commercial joint ventures and employment arrangements that may be struck with the cable company using the local origination channels, and leased access channel possibilities.

Local Cable Program Development, Distribution, and Feedback Possibilities

The access channels offer a distribution outlet for independently produced programs, with audiences at least as large as that of a media art center screening. Perhaps more importantly, these channels offer the producer a rare opportunity to ascertain audience reactions to their work. This can be very useful for a variety of reasons, and accomplished by a number of feedback mechanisms.

For example, if a producer is working on developing the most effective program possible, or one with the most impact, audience evaluations obtained may be instrumental in incorporating the needed changes to achieve those ends. In essence, the producer has a chance to test his or her work "in progress." Obtaining audience feedback information can also be helpful in determining which larger distribution markets may be most appropriate.

Producers may gather such audience feedback by cablecasting their program as part of a live studio program (which, unlike broadcast television, are frequent on access channels). The live program format could be a call-in show that allows the audience to phone the show host and air their opinions of the material. If the producer is present on such a program, usually more audience feedback can be garnered. In another situation a producer may schedule the showing of his/her program and notify all the groups and organizations that may be concerned with the program's content. For example, if the producer has made a documentary portrait of a handicapped individual, all groups who may be concerned about that affliction could be notified by postcard or phone of the show's scheduled cablecasting date. Follow-up discussion groups that meet after the cablecasting could be organized to stimulate more feedback, and more interest in the work (which could lead to more support, and contacts, and production and/or distribution funding). Producers could also make use of the newer systems' bi-directional capabilities for immediate visual and aural responses to their programs. While only a few systems have total two-way or bi-directional capability, most of the systems built since 1979 have bi-directional institutional networks or loops that tie together several community buildings (such as recreation centers, libraries, schools, civic centers, etc.). The discussion grups mentioned could meet in these locations for a live presentation of their thoughts about the program (which may or may not be cablecast). Some producers have experimented with tying microcomputers into the feedback process. This is accomplished by placing a microcomputer with a telephone coupler in the studio site, and then having members of the audience, who have microcomputers, place a call to the computer after the show, and interact with a predesigned computer program. In this way very specific, calculated, and documented reactions can be obtained (which can, depending on the computer program, allow for flexible, interactive responses).

The same bi-directional technology can also be used by video and performing artists for creative experimentation purposes. This technology has seldom been put to artistic uses, although on October 22, 1982 a live, national, interactive experiment called the "Artist and Television" teleconference, was held in Iowa City, Iowa, that linked Iowa City with New York City and Los Angeles. This teleconference began to fuse artists' work with television and to illustrate the many formats, contexts, scenarios, and applications that may be possible, both on a local and a national scale. With the positive responses to this event generated from the art world and the institutions involved, funding for such proposals may be forthcoming. Another experiment with local cable is being tried in Massachusetts. The Massachusetts Council on the Arts and Humanities has issued a Request for Proposals which center on the premise of developing partnerships between producers, non-profit cultural institutions, and cable companies. The Arts and Humanities funds will enable producers and cultural groups to devise programs that cable operators can then cablecast. In part, it is hoped the projects funded will illustrate how programming produced by the groups involved can benefit all groups, and thus become an integral part of doing business in the community. This may motivate the cable companies to contribute the cash and other kinds of support necessary to keep the projects on an ongoing basis.

Local Cable Commercial Joint Venture

Independent producers approaching cable company business people for support or joint venture agreements should be ready to quantify the results of their programs. In order to justify the company's capital risk and other forms of participation inherent in such joint ventures, some way of projecting the program's potential popularity is important. More times than not these business people are going to be very interested in seeing information showing them that your program can obtain a significant share of the local viewership. As far as the company is concerned, this interest usually overrides how well produced or aesthetically pleasing the program might be. If you've developed the program locally, and gathered feedback or survey results as discussed earlier, this may help considerably in attracting the company's interest. If you have no such documentation, and still manage to obtain company support, be sure to retain some of your budget for publicizing the show or the series, surveying the results, and getting the numbers, quantification, and/or justification for your program.

Producers may also make arrangements with the cable company in which revenues from commercials "framing" their programs are shared. Whether it is more appropriate for the producer or the cable marketing department to solicit and do the sales work to get the backing for these commercials will vary from situation to situation (see Chapter Seven, "Cable Advertising" for more information on advertising possibilities). Once it is shown a program will have a viewership, commercials sponsored by local businesses become a real possibility. Some cable companies have not yet entered the area of producing programs or selling commercials locally. Some of these companies will welcome the resources of the producer in the forms of both programming and a commercial sales contract, because the local revenue and viewership generated work very much to their advantage. Other companies are moving into local ad sales aggressively and may be looking for exciting, quality programming in which to sell time. In either case, producer and operator can share the revenue from the commercial sales. Some agreements may or may not allow for up front production funds, but certainly remuneration for costs, time, and effort on the producer's behalf should be accounted for in the negotiations.

Another means of securing revenue from local cablecasts is working with an advertising agency or someone in the area who is in the business of selling local availabilities, or "avails," (local commercial insert spots) on national cable programming being carried by the cable company. Here the producer may do the sales work and/or produce the commercials to be inserted.

If it is not within the producer's experience or financial capabilities to initiate some of the preceeding endeavors, there are other partial employment possibilities. Independent producers can serve as free lance crew on the production of commercials for use in local cable avails and programming spots. While these jobs usually go to those production companies with considerable technical expertise, it is not impossible for the local producer to get a contract, particularly if he or she can undercut the prices frequently caused by the large overhead of a commercial house—and produce an adequate product. If an independent producer has a good relationship with the local cable operator, it is possible to approach him or her with an offer to produce commercials (for local avails or other programs) which the local cable programming staff are simply too busy to do. This enables the cable operator to continue to offer those programming services and net associated production revenues that may otherwise be lost (a number of local avails are never used simply because the local company is not set up to do the production or sales work necessary to make use of these chances).

While the preceding joint venture and employment arrangements make use of the local origination channel, the leased access channel is another means of accomplishing commercially oriented projects.

Local Leased Access Cable Opportunities

The use of leased access channels for distributing locally produced programs and generating ad revenues is still in the very initial stages of exploration. In a leased access arrangement, the leasee simply pays for the channel time and, in return, cablecasts whatever he or she chooses. In many such instances, the leasee, or producer, involved in this arrangement produces both the programming and the commercials, and is totally responsible for the success or failure, profit or loss, of the operation. The channel may be leased for one hour a week, or for several hours a day. The cable industry has been somewhat resistant to the idea of leasing time on an unlimited basis. Such agreements put them in the role of a common carrier, a role which, if generally accepted by the cable industry, would leave them no control over the programming distributed over their systems (the issue of cable companies as common carriers or electronic publishers is now being debated in Congress). However, many operators will lease, or undertake some contractual arrangement, on a limited basis, subject to periodic review and renewal. If a producer has some form of a lease arrangement, he or she can retain all the revenues from sales of ads on their programs. Unfortunately the success of such projects has in some cases contributed to the cable companies' hesitancy in assisting leased access channels to develop, because the leasee can be competing with the cable company for the same local advertising dollars. In any case, the producers with entrepreneurial talent who manage to obtain leased channel use may find this an extremely lucrative source of income.

THE REGIONAL CABLE MARKET

Regional interconnection of cable systems is still a fairly new concept, yet there are already a few examples of such systems being used creatively.

Mr. Bob Williams, of New England CableRep, one company exploring the use of regional cable markets, works with ad agencies and businesses that have an interest in advertising regionally. The company sells commercials and avails, and places them on regional cable systems that reach a larger market area than any one local cable system. Mr. Williams is also interested in developing local programming for regional use in which commercials can also be placed. Two programs have already been developed in this way, including a cooking show and a quiz program.

Lobster Associates, Inc., is a production and distribution company in Brookline, Massachusetts, expressly formed with the belief that money can be made by producing and distributing programming for a regional level (the Boston area). Lobster Associates believes they can shape the best of local programming into regionally attractive fare and also originate new program ventures of their own.

Lobster Associates' progress, like that of similar companies forming around the country, is hampered only by the lack of total numbers of cable subscribers in the region, which should change as more cable systems become interconnected. Although the commercial success of such projects is as yet uncertain, the potential is there for supporting and developing different kinds of programming, and making money in the process.

THE NATIONAL CABLE MARKET

The advent of satellite delivered programming to cable systems has offered increased opportunity for employment to producers, and additional markets for the distribution of independently produced work.

Independent filmmakers Mr. James Brown and Mr. George Stoney made a sixteen millimeter documentary called "Wasn't That A Time," about a folk singing team known as The Weavers. To produce this show, they received some funding from the Bravo national programming service. Mr. Stevenson Palfi, an independent videomaker based in New Orleans, made a videotape about three black musicians entitled "Musicians Almost Never Play Together" and sold the cable rights to CBS Cable. Programs with musical content seem to be of special interest to many national cable services.

HBO frequently involves itself in a project from the outset, according to Ellen Rubin, Director of Public Relations for this service. One recent example is a show called "Catch a Rising Star's Tenth Anniversary," a videotape of the tenth anniversary celebration of a New York nightclub which has given many performers, such as Robin Williams, their start. Rising Star Video, a production company owned by the nightclub, and HBO worked out a contract to produce and distribute the program.

Shorts and filler or interstitial programming, which national pay cable services use between features, may be better received than most other program formats. Amy Devine, Director of Short Acquisitions at Spotlight, a national pay cable service based in Santa Monica, California, reported that their service has acquired four hundred shorts in the two years they have been in operation. Shorts purchased vary from one to forty-five minutes in length. The most frequent shorts purchased were one to ten minutes in length. Most shorts purchased were short dramatic pieces, comedy sketches, and video music pieces. These shorts were bought, for the most part, from Independent Cinema Artists and Producers (ICAP), Phoenix Films, and Pyramid, at fees varying from three hundred to seven hundred and fifty dollars. This subject is further developed in Chapter Six, "Making the Connection: Distributing Original Programming."

Another example of programs developed specifically for the pay entertainment market is the series of original productions performed by the Children's Theatre Company of Minneapolis. The productions are based on the original scripts of "The Marvelous Land of OZ," "Puss in Boots," and "Alice in Wonderland." Mr. Richard Carey, an independent producer with a background in documentary film, managed to produce the series by investing some of his own money, finding other investors, and reaching a distribution agreement with MCA-Universal. The programs will first be seen on subscription television (STV) services, such as On-TV and Starcase, and will also be distributed on video cassette and videodisc. The shows are being withheld from pay cable until MCA-Universal can sell them as a series package.

Producers like Mr. Brown and Mr. Carey agree that if you can break even on the first license sale, or the first airing of a program, you are doing well. It is not until the secondary or tertiary market that it is possible to realize profits on the product. Mr. Stoney points out that frequently you do not make back your production costs if your work is acquired after it is finished. Unfortunately, obtaining development and pre-production funds can be a very difficult matter for many producers, particularly for those who are not well known.

While there are successful examples of producing for, or with, a national cable service such as those cited earlier, it should be noted that these are, for the most part, exceptional cases. That is, a close investigation of whom the pay cable services are usually willing to give development and pre-production funds reveals an unwillingness to risk working with newcomers, or producers without national reputations. In addition, recent developments in the cable industry indicate an increasingly conservative approach to program funding and distributing decisions. Executives in decision-making positions at the major cable services such as HBO and Showtime frequently have a background in broadcast television and make their choices based on knowledge developed there. They are cautious about purchasing certain kinds of programming, because they do not believe the viewing audience wants to look at programming that is non-traditional or experimental, including documentaries (in essence, anything other than features or concerts for most of the pay services).

So, while the number of national cable services has multiplied, and the demand for programming has increased to fill the growing number of cable channels, it does not necessarily follow that producers will find it easier to find development and production funding for the work of their choice, nor for the national distribution of their work. However, some producers have successfully funded, produced, and distributed programs at this level, indicating there are opportunities. The successes of these producers bear closer scrutiny, as do some other cable programming considerations and strategies, that might enable other producers to make use of the existing and potential local, regional and national programming opportunities.

CONSIDERATIONS AND STRATEGIES IN CREATING, PRODUCING, AND DISTRIBUTING CABLE PROGRAMMING

There are many considerations a producer may take into account when planning a program for cable. Some of these considerations can be applied generally to all program production for local, regional, and national markets. For instance, target audiences and program goals should be factors determined in the conceptual or planning stages of all productions. Other basic considerations involved in all programming are decisions about the quality of programming to be produced. The phrase quality of programming has as many connotations as the word television. Most simply stated, quality of programming can be assessed by the following production values: technical, format, aesthetic, and content. Technical values include such considerations as whether to shoot the program with one-half inch VHS or Betamax, three-quarter inch, or one inch video equipment, or whether broadcast standards are, or need to be met. Documentaries, magazine shows, and news programs are examples of different formats. Aesthetics refers to such factors as the lighting, camera angles, and special effects used to enhance the program. Content refers to the substance or messages being communicated. The application of each of these production values, how they work in combination, how they relate to the program's goals, and the target audience response, are the factors upon which the quality of the program can be assessed. Innovation in the use of any of these values (applications beyond the relative traditional uses, such as special artistic enhancement) and/or in the use of the technological delivery system being used (i.e., such as interactive uses of cable, finding new applications for institutional networks, etc.) adds another dimension to program quality and its assessment.

Some of these values are determined when decisions are reached as to target audiences and program goals. For example, if a community access producer wants to make a talk show about a local nuclear energy plant, and the intent of the program is to give local residents the opportunity to learn as much as possible about that plant, several production values become apparent. Since the program is intended for local cablecasting, shooting on either one-half inch or three-quarter inch video equipment is sufficient. Since the show's goal is to communicate information, some aesthetic considerations such as artistic special effects or dramatic lighting may be dropped in lieu of putting as much energy as possible into the content of the program.

On the other hand, if a producer is trying to make a living at producing and selling programming, other production values will take priority. It is in this producer's interest to: use the best equipment obtainable, and if possible a camera with at least four hundred lines of resolution; maximize and tailor each of the other production values according to the market or service targeted; make a first generation master for all dubbing and put the master in a safe; include an engineer from day one (for new producers or those financially strapped, cultivating a student engineer from a local college or university is one option). A good, and sympathetic, engineer can often bring footage that would otherwise be unusable up to broadcast specifications, which is all that is necessary for satellite delivery. Also, make connections with a good distributor who will help you sell your work. Each of these suggestions leads to a product that gives the producer the most options possible. Today it is best, when possible, to produce a program for all available markets, including: exhibitions in art centers or museums; local, regional, and national basic and pay cable; broadcast; foreign; and videodisc. Each of these suggestions can help the producer achieve this end.

Other cable programming considerations and strategies relate only to national markets. For example, although Mr. Brown and Mr. Stoney's program based on the musical group is an exceptional documentary work, it is less for this reason than the nature of the material incorporated into the program that it was purchased. That is, the content of the program was music, and presently the market for music programming is the most salable, and one of the largest. Documentaries, on the other hand, are next to impossible to sell to most national cable services. This example indicates several common characteristics of national cable services which very quickly presents the producer with several inescapable choices.

The national cable services have, to a large degree, emulated the networks. Most services carry much the same fare as broadcast television, such as movies, sports, news, and entertainment. Such fare is the product of broadcast-trained executives in those services that have recognized and reinforced established program formats, content, and patterns. Such formats, content, and patterns have, more than any others, shown the potential for consistent and increasingly larger viewership. That, in national cable terms, is the game and goal. The businesses running the national cable services are in the business of making money. Documentaries, experimental, artistic, and other nontraditional or innovative programming is, therefore, unlikely to be warmly received by such services; until it can be shown that such programming will have a substantial viewership, and even more preferable, broaden their subscriber base.

For those independent producers who make programs for exhibition in art centers, museums or on local access channels, and, when lucky, get a one-time showing on PBS, such considerations as attracting an audience in the millions, or broadening a service's subscriber base, seldom enter their conceptual framework or their production planning. These matters need to be taken into consideration if the producer's goal is national sales or distribution.

It is at this point that many choices are presented to producers. For those producers who wish to make the most salable products to the national services, or gain employment with them, a close examination of the services you wish to sell to, or be employed with, is to your benefit. Learn as much as possible by observing their programming formats, content, and patterns. For independents, presently shorts and music pieces seem to be the most salable program material.

Because of different interests and motivations, other producers will choose to concentrate on creating what they consider different or better quality programming, consciously trying to avoid the traditional fare of the national cable services. If the producer's intent is to innovate, or communicate artistic, aesthetic or content-oriented material that is presently not marketable, perhaps working on the local access, local origination, leased access, or regional level is most appropriate. Programming of this nature, its viability and funding, can be developed and experimented with most easily on these levels, until strategies and circumstances allow for national distribution. If national sales or distribution is a goal of the producer, it should be recognized that only a sales strategy that incorporates the benefits to those services in their terms (i.e., the programming will attract a large viewership, broaden their subscriber base, etc.) is likely to succeed.

Some examples of different, better quality, or innovative programs have been developed on local access channels and moved to other markets; and some program innovations have turned up on the national cable services. Mr. Gary Knowles started a number of shows on the Madison, Wisconsin access channel. One of these, eventually called "Live On Six," was moved to the local origination channel, and then to the Madison NBC broadcast station. Mr. Leon Varjian started the "Vern and Evelyn Show" (also on the Madison access channel), which was taken on by the CBS broadcast station in Madison. Home Box Office (HBO), a national pay service, has illustrated some innovative programming, such as the "Yesteryear" series. This series makes use of chroma key, good lighting, and sense of scale to project its host, Mr. Dick Cavett, into scenes in which he appears to be a participant in 1930s documentary footage. CBS Cable, before its demise, presented "Signature", a show that turned the age old format of the talk show into exciting television. "Signature" featured an active, involved host who was seen only as a silouhette. Cameras gently framed guests' faces with closeups rarely seen on television. There were no cuts, only dissolves from camera to camera that never left the subject's face. While these examples do not represent the calibre of attitude shift necessary before national services will accept dramatically different kinds and forms of programming, it is indicative of some national service's willingness to experiment, innovate, and consider more artistic or aesthetic values.

CONCLUSION

Many opportunities exist for producing and distributing original programming for local, regional, and national cable markets. These opportunities take many forms, and are different for each market. Local channels offer a great deal of resources, flexibility, and opportunities for exploration and experimentation. Regional markets are only beginning to be explored, yet lucrative possibilities are already foreseen. Independent producers in both of these areas may be involved in a number of possible scenarios with access, local origination, leased access, or regional cable channels, which may take the form of community producing, independent program development, free lance work, producing commercials, or joint venture arrangements.

The national cable services, while offering an expanding number of channels, are much more difficult to produce for. Yet again, producing and employment opportunities do exist for those interested in and capable of tailoring their programming to these services' needs and interests, or proving that the programming offered fits those needs. When producing and distributing original programming for any cable market, many considerations and strategies need to be faced and decided upon, according to each producer's interests, motivations, and goals. The resulting decisions become the blueprints to help determine what producers can produce and distribute for which markets.

Whether the producer's interest is community programming, commercial endeavors, employment, or program experimentation and innovation, on one level or another, these opportunities exist in cable.

RESOURCES

1. Barczyk, Fred, Producer, WGBH, 125 Western Avenue, Allston, MA 02134; 617-492-2777.
2. Judson, Lisa, Affiliate Relations, ABC ARTS, 825 7th Avenue, New York, NY 10019; 212-887-7777.
3. Leacock, Richard, Professor of Film, Massachusetts Institute for Technology, Building N51, Room 115, 77 Massachusetts Avenue, Cambridge, MA 02139; 617-253-1000.
4. Liroff, David, Producer, WGBH, 125 Western Avenue, Allston, MA 02134; 617-492-2777.
5. Logan, Jennifer, Home Theatre Network, 465 Congress Street, Portland, ME 04101; 207-774-0300.
6. Massachusetts Council on the Arts and Humanities, 1 Ashburton Place, Boston, MA. (Cable Project Coordinator).
7. Morgan, Kitty, Independent Cinema Artists and Producers (ICAP), 625 Broadway, New York, NY 10012; 212-533-9180.
8. Rubin, Ellen, Home Box Office, Time and Life Center, Rockefeller Building, New York, NY 10020; 212-484-1000.
9. Stoney, George C., Professor of Film and Television, New York University, Alternate Media Center, 725 Broadway, 4th Floor, New York, NY 10003; 212-598-2852.
10. Williams, Bob, New England CableRep, 31 Fairfield Street, Boston, MA 02116; 617-267-8582.

Fred Johnson and Wm. Drew Shaffer

4. Independents and Access: The Status, the Predicament, and the Promise

Three common needs facing independent producers and others interested in creating original programming for cable television are securing the necessary production resources to make their programming, finding ways of distributing that programming, and figuring out how to get paid for their work. Following are suggestions on how independent producers can fulfill some of these needs by making use of what is offered by the many sophisticatedly equipped cable television community access centers now developing across the country. In addition,

---

Fred Johnson has been an independent producer and cable consultant since 1975. He is employed as the Access Coordinator for a regional consortium called the Four C's in Campbell County, Kentucky.

Wm. Drew Shaffer is the Broadband Telecommunications Specialist for the City of Iowa City, Iowa. Mr. Shaffer is a producer of educational, fictional, documentary, and commercial work for such institutions as NEH, NCAT, and the Iowa Governor's Safety Council. He has also been employed as a video and cable consultant since 1972 and has taught at the University of Iowa.

Copyright 1983 by National Federation of Local Cable Programmers (NFLCP).

some of the problems independent producers are likely to encounter when first dealing with these access entities are outlined and ways those problems may be resolved to the satisfaction of both independent producers and access centers are also suggested.

THE STATUS: WHAT THE ACCESS CENTERS OFFER

Since 1979 community access packages have been offered to many cities as part of franchising contractual agreements by cable companies. Many of these packages have included large amounts of state-of-the-art production equipment for access purposes; from one to dozens of community access or institutional channels; and production grants of up to hundreds of thousands of dollars to local individuals and groups to create programming germane to the particular area. Most of the access center services—the workshops, videotape, production equipment, and channel time—are offered at minimal or no cost. The access channels alone are of great benefit to the independent producer. These channels open another outlet to exhibit independents' work to the public; to get their work known; and to build audience, community, and prospective clients' confidence in their capabilities. Sometimes, when politically unpopular material is involved, the access channel is the only outlet for an independent producer's work. Many granting agencies and foundations are now including in their application forms the stipulation that media work produced by their funds must be distributed as widely as possible. Sometimes use of an access channel is specifically written in as a distribution requirement. Many of these centers and access channels are rapidly becoming the matrix within which all production on the local level occurs. Inherently then, it would seem that within these access packages is a plethora of answers to many of the independent producer's identified needs—perhaps; perhaps not. A more careful examination of individual access situations may reveal a number of problems.

THE PREDICAMENT

While access centers offer a great many resources and services to independents and the general public alike, there may be a number of curious, confusing, or troublesome conditions evident upon approaching any individual access center. That is, any access center an independent producer approaches is likely to be not only very different from any other entity that producer has dealt with before; different procedurally, operationally, and philosophically, but also different in many ways from other access centers. So, the predicament faced by independent producers is to know the kinds of problems they may have to deal with when approaching an access center, and then how to deal with those problems.

How Access Centers May Vary

Two scenarios depicting access situations in different localities illustrate some of the problems. In the first hypothetical case, you are an independent producer with some notable credits and experience entering an access center for the first time. You are shown two three-quarter inch portapaks, one three-quarter inch editing system, and one studio. You are asked if you are a resident of the county. If not, you can't use the equipment. You are told everyone is treated on a first come, first served basis; that everyone who uses the equipment must take a workshop; that you can have twenty hours of equipment use per month; and that the programs you produce must have some relevancy to the community. The implication is made that your program, in terms of relative importance, fits somewhere between a talkshow for senior citizens, a videotape made by someone who had never held a camera before, and the local Jaycees' coverage of summer softball. You are then told no payment is involved for any program made with the access equipment. The programs you make must be shown on the access channel five times, and the copyrights to such programs belong to the cable company. This last stipulation alone is usually enough to stop many independent producers from proceeding further. If it doesn't, and one manages to produce a program and get it shown on the access channel, the producer's own "broadcast-type" expectations from the access cablecast may be the final disappointment that will end any future cooperative efforts.

In the second scenario you are an experienced independent producer who walks into an access center and are given a tour of the variety of state-of-the-art equipment which includes two dozen three-quarter inch portapaks, six one-half inch VHS portapaks, four three-quarter inch editing systems, two fully-equipped mobile vans and two studios. You are told this equipment is available to everyone who is qualified to use it, and that you can take workshops, if you need them, free of charge. You can use the equipment as much as forty hours per month at no charge. Up to sixty minutes of tape can be checked out for as long as two months. You can request, again at no charge, up to twenty hours of access channel time per month to exhibit the work produced. You must show the program at least once on the access channel but you will own the copyright and production grants are available to help recoup expenses for making the program.

These two scenarios illustrate some of the more common variations and problems exhibited by access centers. For instance, there is production equipment available, but there is never enough equipment. Because of this, limitations are imposed on the number of hours the equipment can be used by any individual. Sometimes geographical boundaries are drawn to limit who can use this equipment (i.e. the user has to reside in the same city or county in which the center is located). In terms of procedures and operations as well, centers vary considerably. While one center may require even experienced producers to take workshops before using any equipment, another center will allow test-outs. In another instance one center will require one showing of whatever program is produced with the access equipment, and another may require five or more showings. Certainly a major issue of variance from center to center is copyright ownership. While most centers allow the producer to retain all copyrights, some may not. Access centers show almost no variation though in their intent to teach anyone who is interested how to make television programs, and in that there is no direct payment for programming produced. Philosophically it is access' purpose to create a free flow of information and programming.

Without knowing the reasons leading to the problems in the preceding scenarios and how to deal with those reasons and problems, the independent producer involved is liable to be disgruntled or insulted, and not initiate further contact. This is a great loss to both the access center, which does need and want good programming, and the producer, who could find the access center just the resource s/he needs. How do such wide discrepancies in access center operations exist? In order to avoid, or help change the problem situations exhibited in these scenarios, and so that both the independent producer and the access center can get the most from their relationship, an understanding of why and how access centers operate as they do is imperative.

Why the Access Centers Vary

The most important thing to remember about access centers is that they vary in operation from city to city as much as do state license plate colors—and for several reasons.

Access started as a set of ideas taught at the Alternate Media Center (AMC) at New York University in the early 1970s (although there were a few isolated instances across the country, such as in East Lansing, Michigan, where access centers were started without the AMC's guidance). The number of students who received those ideas was small; nine in 1973. They took these to various cities across the country to test them. Many of the ideas they promoted and worked to establish were new, yet there was a ring of truth and excitement to them that many recognized. These ideas were tantamount to affixing a decentralized, democratic, and flexible media function and structure to the growing cable industry and were to be specifically implanted in the local communities. This was no small task, particularly without any major institutional backing, and there wasn't much of that until the late 1970s. In 1977 the National Federation of Local Cable Programmers (NFLCP) was formed to help cities cope with the onslaught of cable franchising, by ensuring that the cities knew the alternatives and opportunities that could be negotiated. These include: community access equipment packages; access channels; institutional networks; production grants; and franchise fees. Still, the number of companies franchising was so great that far more cities were franchised than the NFLCP or the Cable Television Information Center (CTIC, a similarly motivated and involved organization) could assist. Many cities have had no guidelines or assistance from such knowledgeable sources and negotiated with the companies from a variety of positions and ideas which resulted in a full spectrum of access realities. On one end of this spectrum are cities with no access channels and no access equipment. On the other end are cities with more equipment and channels than they'll know what to do with for years. So, in 1982, different cities exhibit almost every conceivable variation on the access theme. This is partially because the business of establishing such a novel set of ideas and structures is quite complicated and may take many years to develop; partly due to the lack or variety of guidance; and finally because there is no one perfect blueprint to community access. By its nature it must develop as a response to each community's needs. Therefore its composition and operation will, to some degree, vary from city to city. However, because there are usually certain common elements in each access center and community, three actions can be taken to resolve most of the problems or incongruities that exist.

WORKING TOWARD A RESOLUTION: THE PROMISE

The first step any independent or citizen can take to resolve problems such as those described here is to find out the operating rules or guidelines of the center and who is responsible for creating and overseeing those rules. The second is to learn at least as well as those who created the center and rules, the fundamental presuppositions, principles, and goals of access (upon which the center and rules were founded) so that you can make the system work for you. Finally, by understanding and communicating the similarities in what you as an independent producer want and need and what access' aims are, you can arrange many cooperative alternatives to benefit yourself and the access center.

The Operating Rules

There are no federal guidelines that can be applied to access centers. The only federal standards applying to the access channels are that they are to be used for non-profit purposes, and are not to carry any lotteries, political endorsements, commercials, obscenity, or pornography (which in reality is left to "community standards" interpretations). Beyond this, the operating rules are open to discussion and negotiation. In some cases the rules for the center and channels are negotiated between the city and the cable company at the time of franchising, and are included in the franchise proposal offered by the cable company. In most cases such details are left to be negotiated later.

You should obtain a copy of the rules (if the city has been involved, from the City Clerk's office) and identify the key players in their creation. Find out who oversees the rules. Usually a cable commission, an advisory body to the City Council, has been set up to monitor such areas. Sometimes a cable television coordinator has been hired by the city to assist in overseeing the company. If problems exist with the rules or the way the center is run, first try to work it out with the people who run the center (i.e., the cable company, a school, library, non-profit organization, or the city). If satisfaction can't be obtained, take the matter to the commission. Oftentimes the rules can be modified or changed, if what you need is seen as being consistent with the community's interests and the basic access principles being applied in your community.

The Principles of Access

The ideal foundation of access is built on several identifiable presuppositions, principles, and goals. Frequently only a few of these are transmitted to (or thought of by) an access center or city when franchises are granted or when rules are drawn up. As an independent producer or interested citizen, knowing these can assist you in developing the system into what it should be and thus getting from it what you need.

Two presuppositions of access are that information (which is synonymous in this context with programming) is a vital resource which cable technology is uniquely capable of distributing, and that everyone should have access to this resource. That there is an inherent value in citizens' understanding visual communications, and in having an opportunity to express their ideas and opinions are also basic suppositions. Access presupposes a backlog of need in the community to communicate visually (for other than commercially oriented purposes), and that the community will "pay" for this programming in the form of time and energy to produce it (everything but the labor is of minimal or no cost). Failing this, the community will pay others, such as independent producers, to produce programming for it.

The goal of access is to create flexible, decentralized, democratic media systems in each community, that allow for that community's communication needs. This involves the notion that access is a common carrier concept, and as such facilitates a free flow of information and programming. Access means locally controlled and autonomous channels which are meant to preserve and protect each community's indigenous populations and unique needs. Access works to enable the creation and distribution of diverse programming. Programming that is information centered as well as attention oriented; is content or message as well as form oriented; allows for a full range of social, humanistic, and artistic applications and learning as well as self-expression; and is by its nature local (made by, for, and about the community). Because of the unique mix of programming, cable, and video resources becoming available, a qualitative change in programming that creates new relationships between the media producers, distributors, and consumers (based in part on dialogue and feedback mechanisms) is possible. This programming is participational and user-controlled—a totally new way of thinking about and using television.

Alternatives that Benefit Both the Independent Producer and Access Center

A number of similarities exist between what many independent producers are trying to accomplish and what access represents. Quite a number of independent producers work to explore the potential of media beyond a commercial emphasis, and use the medium as a form of social-, artistic-, and self-expression. Many other independent producers work to eliminate the discrepancy between the potential of media and its present performance. From such a basis, it is apparent there are many common elements at work for the independent producer and those who represent access. This commonality can be a starting basis for a good working relationship.

The access philosophical intent to create a free flow of information and programming is, in many ways, perhaps the most contrary notion to independent producers. While access functions to teach everyone who wants to learn how to make television programs, this is an ideal; an ideal that, in reality, will never be attained. Not everyone can or wants to learn about television. But access exists to enable as much programming to be made by as many people as possible. All programming is important and is of the same value. Although access emphasizes that no direct payment for programming is made, the centers and channels are excellent entry opportunities and resources. The workshops, equipment, and channels offered are incomparable to what is available from any other media entity in existence. For those independents whose livelihood depends on making programs, a number of possible funding and cooperative arrangements exist.

Some centers allow PBS-style sponsorship of programs, allowing the producers to recoup some of their costs. This can be an excellent way of developing programming, expertise, and clients. Many of the newer centers also offer production grants to support any programming that is relevant to local concerns. Because access channels need programming, independent producers can offer packages of programming on a limited playback basis in return for access to some of the center's resources. Many access centers have difficulty teaching all the people who want to learn about television and production techniques. Independent producers can teach workshops in return for satisfaction of some of their needs. Productions of all kinds are constantly in process at most centers, and production assistance from independent producers can insure a higher quality of programming, and help the access center staff. At the same time the producer can cultivate sponsors of such programs (i.e., community organizations, schools, the city, etc.), who then start to identify the need for such expertise. For example, independent producers can assist a number of community organizations with similar interests to produce programming which serves to meet mutual goals (whereas any one organization may not be able to afford such assistance). Direct funding for producing programs then becomes a good possibility. In some places access centers will allow use of their resources at special rates for non-profit community organizations.

In other situations, where an independent producer plans to produce programming for other than access purposes, some centers have negotiated flat fees (or a percentage of return on whatever proceeds come from the work) in exchange for the use of their resources. The proceeds then revert to support for the access center.

Some producers move from making access programming to developing new communications processes and avenues such as producing call-in talk shows; arranging real-time, live, interactive meetings or conferences between various community entities; developing the bi-directional capabilities and institutional networks of the cable system; and coordinating live satellite links with other cities. Still others move into the commercial opportunities offered by local origination or leased access channels.

CONCLUSION

Access centers and independent producers have a great deal in common and can establish relationships that will support each other. Each is exploring and opening up areas and resources for the other. While access centers operate in certain ways for identifiable reasons which were established for their communities, usually an attempt is made (or should be made) to allow the most number of alternatives and possibilities to exist within a certain set of parameters. By identifying these reasons and parameters, independent producers can better work with access systems or help change them as necessary. No other groups have a better understanding of the possibilities and potential of television, nor are so directed to such ends, as independent producers and those who support access centers. By their activity and efforts, independent producers who do become involved can help to define and shape access systems that will best benefit the communities, access, and themselves—and thus help to realize the promises projected onto the medium, and those purported by it.

RESOURCES

1. Battcock, Gregory, *New Artist Video,* E.P. Dutton, Inc., 1978.
2. Davis, Douglas, and Simmons, Allison, *The New Television,* MIT Press, Inc., 1978.
3. Enzensberger, Hans Magnus, *The Consciousness Industry,* Seabury Press, Inc., 1974.

4. Gerbner, George, Gross, Larry, and Melody, William H., *Communications Technology and Social Policy*, John Wiley and Sons, Inc., 1973.
5. Gillespie, Gilbert, *Public Access Cable Television in the United States and Canada*, Praeger Publishers, Inc., 1975.
6. Hollowell, Mary Louise, *The Cable/Broadband Communications Book, Volume 2*, Communications Press, Inc., 1982.
7. Illich, Ivan, *Shadow Work*, Marion Boyars, Inc., 1981.
8. Schiller, Herbert I., *The Mind Managers*, Beacon Press, 1973.
9. Seiden, Martin H., *Cable Television U.S.A.*, Praeger Publishers, Inc., 1972.
10. Williner, Alfred, Milliard, Guy, and Ganty, Alex, *Videology and Utopic*, Routledge and Kegan Paul LTD, 1976.
11. Youngblood, Gene, "The Mass Media and The Future of Desire," *Co-Evolution Quarterly*, Winter 77/78, Box 428, Sausalito, California 94965.

ORGANIZATIONS

1. Association of Independent Video and Filmmakers (AIVF), 625 Broady, New York, NY 10012 (212) 473-3400.
2. Cable 10, 222 Clinton Street, Franfort, KY 40601; (502) 227-4480.
3. Fayetteville Open Channel, 309 B West Dickson, Fayetteville, AR 72701; (501) 521-9870.
4. Kentucky Independent Video and Film, Inc., P.O. Box 141, Frankfort, KY 40602; (606) 233-7637.
5. Marin Community Video, 61 Tamal Vista, Corte Madera, CA 94925; (415) 924-7370.
6. National Federation of Local Cable Programmers (NFLCP), 906 Pennsylvania Avenue, SE, Washington, D.C. 20003; (202) 544-7272.
7. NOVAC, 2010 Magazine Street, New Orleans, LA 70130; (504) 524-8626.
8. University Community Video, 425 Ontario, SE, Minneapolis, MN 55414; (612) 376-3333.

Ernest T. Sanchez, Esq.

5. Basic Legal Planning for the Producer of Original Programming, or I Wish I had Thought of that Before We Started Production

Film and video program producers don't like to think of what they create as being pieces of property. They point to the vivid imagination, superb technical skills, and endless patience that go into their unique final product. Perhaps because a very special kind of magic is involved in this process, producers sometimes lose sight of the fact that their creations share certain legal characteristics with more ordinary kinds of property. It represents no disrespect to state that a program or film can be both a work of art and a piece of property simultaneously. A practical appreciation of this concept is essential for the producer who wants to avoid financial and legal disaster, and who aspires to make a good living from these creations.

There are dozens of legal questions that have to be dealt with in many aspects of the planning, producing, and distributing stages of a program. Because no two productions are exactly the same from a legal standpoint, there can be no way of knowing what exact legal questions may arise, and therefore no substitute for knowledgeable legal counsel. However, there are a number of key legal questions a

---

Ernest T. Sanchez, Esq. is a partner in the Washington, D.C. law firm of Liberman, Sanchez and Bentley. Mr. Sanchez has served as General Counsel to National Public Radio, and an attorney for the Corporation for Public Broadcasting, and is immediate past Chairman of the Editorial Advisory Board of the Federal Communications Law Journal.

Copyright 1983 by Liberman, Sanchez and Bentley.

producer should consider in the planning stages of a production, involving ownership, source material, and talent and personnel agreements. This chapter concentrates on some of these key legal questions and issues in the planning stage of a production.

WHO WILL OWN PART OR ALL OF THE COMPLETED PRODUCTION?

Ownership of a completed production is almost always directly related to how the production is being financed and how the financial risks of the production are being shared by the parties involved.

If the producer is a paid employee, the answer is usually that the employer owns everything created by the employee for all purposes in perpetuity. While other arrangements are sometimes negotiable between employer and employee, the leverage available to the employee is usually minimal in this type of situation. An exception to this would be one in which the employer fears losing a highly skilled employee unless some of the production's property rights are shared with the employee.

Shared Ownership Arrangements

If the producer does not need anyone else's money to finance any aspect of the production, then the producer will be the sole owner of the final production. Unfortunately, few producers find themselves in such a position. Consequently a variety of ownership arrangements are born. The nature of these arrangements is dependent upon a number of factors, including how and from whom the producer receives support for a production.

Commissioned Work

When a producer has been commissioned to create a work there may be considerable room for negotiation of some ownership rights, since the commissioning person or organization might only be interested in utilizing the final work for some relatively narrow purpose or for a limited duration of time.

For example, a commissioning organization may be primarily interested in funding a project for use in conjunction with a particular event, such as an anniversary. The producer might then be in a position to retain ownership of the completed work, while the commissioning organization would receive certain narrow rights to utilize the work for a limited time and for specific purposes. If the producer has received minimal compensation for such a project, the sharing in production ownership may be a negotiable item for additional recompense.

Another example of a commissioned project ownership arrangement is when a producer is a grant recipient. Usually a patron-of-the-arts grantor will not expect outright ownership of the completed project. Instead, the grantor might expect to share in the project revenues in some fashion. For example, grants from the National Endowment for the Humanities (NEH) and the Corporation for Public Broadcasting (CPB) generally require certain ancillary revenues to be shared by the producer and the grantor. CPB's standard contractual language, for example, states that: "Any net proceeds resulting from the subsidiary or ancillary use of any program produced with CPB support hereunder shall be shared with CPB in proportion to CPB's contribution to the total production costs of the program, up to a maximum of fifty percent of the net proceeds." Subsidiary or ancillary rights are defined by CPB as including, ". . . cable television or origination rights, subsequent commercial or foreign television broadcast rights, merchandising licenses, cassette distribution, videotape, videodisc, audiovisual or other derivative rights . . ."

Co-Productions

Another ownership arrangement involves a co-production agreement between a producer and another party. These arrangements usually take the form of the second party providing the project with money, "in kind" contributions, or a combination of the two. Again the division of ownership rights between the producer and the funding party must be considered and negotiated, in addition to determining how to avoid or resolve questions about editorial, artistic, and production standards.

Co-productions often feature an "active" producer and one or more "passive" producers. These are relative terms suggesting that it is possible to structure a co-production arrangement in ways that leave effective production control in the hands of the active producer. The passive producer is the party providing the funding, or a portion of it. Sometimes passive producers are motivated to enter into an international co-production as a means of securing, in advance, some category of exclusive rights to exploit a production in a specific geographic territory.

Co-productions often represent an effective way of pooling resources and spreading the risk involved in undertaking an expensive production. This joining of resources may help the co-producers create a better program than would otherwise be possible for the total number of dollars each has available, and may actually reduce the cost borne by each party to produce the desired program.

There are many ways a co-production may be formed. Let's say Producer Smith has an exclusive option on a literary property but not enough money to finance the production. Producer Brown has the needed money and becomes a co-producer by providing the necessary funding. The two parties then negotiate an agreement to divide the rights or revenues from the co-produced program in some mutually satisfactory manner.

In another example, Producer Smith has developed a concept and shooting script for a documentary but has neither the necessary technical equipment nor the money to rent it. Producer Brown is in a position to provide all the necessary technical equipment, and the two parties become co-producers.

Or perhaps, Producer Smith wants to create a program about a particular immigrant group but does not have sufficient resources or expertise. Producer Smith seeks out a broadcast organization in the home country of the immigrant group and they agree to pool their production resources. They divide the program rights so that the foreign broadcaster, a co-producer, receives the program rights for its country, and the right to share in revenues from other uses of the finished program.

As can be seen from these examples, everything in the area of ownership is negotiable. As a practical matter, the ability to claim partial or full ownership in the final program is almost always directly related to the desires of the major funder. For this reason it is extremely important that the producer analyze the motivation of the funder(s) before negotiation in the planning stage of a production. Usually the major funder of a project is able to dictate in whose name the program will be copyrighted, how licensing of the program will be handled, and how revenues from the program will be divided. In some instances, for lack of forethought and such crucial analysis and negotiation, producers desperate for completion funds have been known to give substantial or even majority ownership of the project to a funder who contributed only a small percentage of the total project funds.

On the other hand, some producers have successfully negotiated having other people pay for an entire production, while the producer retained ownership of the program. Sometimes granting organizations have funded projects and negotiated for no more than a credit on the completed project.

The key is to know what the funder wants or what motivates the funder, and to negotiate accordingly, knowing what you as a producer want and need. While many producers are capable of successfully negotiating a beneficial ownership arrangement on their own, a lawyer conversant in the intricacies of program financing can help a producer identify the legal structure that offers the best means of securing and protecting the producer's desired ownership interest in the production. Whether a lawyer is involved or not, this has to be done in advance of starting production, and written understanding with all funders should be signed before production begins.

WHAT SOURCE MATERIALS WILL BE USED IN THE PRODUCTION?

There are often a number of source materials needed to create and complete a program. Some of those sources include literary materials such as novels or scripts; original materials like music, costumes, and scenery, and existing materials such as photographs, paintings, and stock footage.

Literary Materials

A production normally will be based on a script adapted from an existing literary work such as a novel, play, or short story, or an original work created especially for the production. If an adaptation of an existing work is being made, the next question is whether the underlying work is copyrighted or is in the public domain. If the work is under current copyright protection, permission to make the adaptation must be secured from the owner with appropriate rights arrangements and fees negotiated. It is important to keep in mind that the copyright owner of a particular work may not necessarily be the person who created the work in the first place. The creator may, for example, have sold the copyright outright or transferred certain interests in it to a publisher.

Frequently, if a production has not yet been funded, the producer will seek an option on the underlying work to ensure its later availability. An option agreement will generally specify that in return for a specific, nonrefundable fee, the work will be exclusively available to the producer for a set period of time. If during the option period the producer begins production, additional fees and royalties, as specified by the agreement, are payable by the producer to the copyright owner. A license agreement for the underlying work should include not only terms relating to fees and the length of time the work may be used, but should also specify what rights of approval, if any, the copyright owner will have over the script based on the original work. In addition, the agreement should specify whether the producer is acquiring the right to make only one specific program or something more extensive, such as a series based on characters, settings, or incidents from the original work. Regardless of whether a particular script is an adaptation or an original work, an agreement with the script writer will also be necessary to set out those rights and fees arrangements.

If the literary work to be used in the production is in the public domain, the producer is free to use the written material as s/he chooses because there is no longer a copyright in effect. A producer can usually determine if a work is in public domain by doing a copyright search. A search can be accomplished by sending a fee (there is a ten dollar per hour charge) and information about the work, such as the date of publication and the names of the author and the publishing house to: Copyright Office, Library of Congress, Washington, D.C. 20559. The copyright office will then determine the copyright status of the specified work. You can do your own copyright search if you desire, although it is long, tedious work. Information explaining how to do your own copyright search is available from the same office at the above address upon request.

Original Materials

Virtually every producer will need to commission original elements or other material into a production at some point. In a major production these commissioned items might include original theme music, scenery, costumes, special effects, or animation. In a smaller production these elements might include titles or still photographs. Another category of commissioned work sometimes overlooked is promotional material prepared for use in conjunction with the production. This would include items such as press kit designs, art work, logos, posters, and teacher's guides. Agreements commissioning these elements should set out not only fees but also provisions related to required approvals, ownership of originals, and re-use rights.

Existing Materials

Many productions, especially documentaries, rely heavily on materials originally prepared for other purposes or for other productions. Such materials commonly include stock footage, excerpts from other productions, still photographs, paintings, music and excerpts from literary works. To the degree that such materials are under current legal protection, permission may have to be sought for their use, and acquisition agreements entered into. Such agreements should always be concluded before the actual utilization of material in the production. The exclusivity that the producer will be granted in the licensed material is an important element of the acquisition agreement, in addition to fees.

WHAT PERSONNEL, TALENT, AND OTHER ARRANGEMENTS NEED TO BE CONSIDERED IN A PRODUCTION?

There are a number of considerations to be made regarding all persons involved in a production. Personnel and talent agreements are two very important considerations to be carefully assessed. In addition, others appearing in the production, such as a non-paid interviewee in a documentary, present additional factors the producer should consider.

Personnel and Talent Agreements

In general, program producers prefer to pay production personnel and talent on a one-time flat fee basis with all production rights for all purposes transferred in perpetuity to the producer. This avoids the complications of paying periodic royalties, and perhaps having to negotiate for additional categories of program usage rights at a later date.

While unionized technical personnel are generally available on a one-time fee basis, this is almost never the case with talent represented by major entertainment industry unions such as the American Federation of Television and Radio Artists (AFTRA), Screen Actors Guild (SAG), Actors Equity Association (AEA), the American Guild of Musical Artists (AGMA), and the American Federation of Musicians (AFM). In general these unions have established standard industry agreements which specify talent fees based on categories of program usage for specific limited lengths of time. With a few narrow exceptions, one-time "buy-outs" for a flat fee are virtually impossible to secure under these agreements.

In addition, each union usually has a wide variety of agreements which apply to specific production categories. For example, the AFM has different agreements which cover musicians providing services in the following: theatrical motion pictures; television movies; syndicated and network television programs; documentary and industrial films; public television; and public radio. Understanding and complying properly with union agreements can be both complex and time consuming. Most well-known professional talent will only be available under the terms of applicable union agreements.

In a union context the producer must consider to what degree the program rights available under each specific agreement are co-extensive with those in other applicable agreements, both in length of time and rights granted. If the program rights for all elements of the production are not co-extensive and do not properly mesh together, the producer's ability to utilize the production will be limited to the narrowest rights available for any individual production element. In other words, it may do the producer very little good to secure and pay the talent for unlimited utilization of their performances if the arrangement with the owner of the underlying copyrighted work only specifies two years of use without an option for renewal. In general the basic rights initially secured should be as parallel as possible to each other, and special care should be taken to make sure that all options for extensions and for additional rights are also parallel. All such agreements with production talent, as with technical personnel, should always be in writing.

Other Persons In the Production

As a rule of thumb, there should be some form of written agreement with every person who receives compensation for appearing in the production. In addition, there are a number of circumstances where a written release is highly desirable from a person who is making an uncompensated appearance, such as a teen-age unwed mother who is interviewed in a documentary. These instances would include, for example, appearances by minors and mental incompetents (the release should ideally be signed by the person appearing in addition to the parent, guardian, or custodian) and "in performance" appearances by entertainers before a paid audience, unless the usage is of the spot news variety. In general, no written permission or release is necessary from persons in studio audiences who by their physical presence have already consented to appear in the production.

OTHER USES OF THE FINISHED PRODUCTION

Sometimes a producer may have several possible uses in mind for the production being created. In some situations it may not be possible to acquire all rights for all purposes in perpetuity for the elements in the completed production. In that case, at a bare minimum, option arrangements should be incorporated in the appropriate individual contracts specifying what additional rights are to be available and under what terms. In particular, specific careful attention should be paid to clarifying whether a given option may be automatically exercised by making a required payment or whether there is some other prerequisite to its exercise. This way, it is clear when and under what circumstances the producer can use the production.

CONCLUSION

The questions and issues presented in this chapter represent some of the considerations a producer should take into account in the planning stages of a production. These are only a starting point from which the producer, and as s/he chooses, experienced production legal counsel, can potentially arrange the best possible conditions for the successful and profitable use of the production. A sympathetic and knowledgeable lawyer may be a cost-effective member of your production team, assisting you through all the legal considerations in each phase of a production. Forgetting or ignoring such criteria as those set out in this chapter until after production is underway can only multiply the legal risks, complications, and costs of trying to sort things out in-progress or after the fact.

RESOURCES

1. Nimmer, Melville B., *Nimmer on Copyright,* Matthew Bender Publishing Co., 1982.
2. Sanchez, Ernest T., "Negotiating the Acquisition and Licensing of Television, Radio, and Cable Programming: Key Legal Considerations," *Current Developments in Television and Radio,* The Practising Institute, New York, NY, 1982, pp. 251-288.
3. Sanchez, Ernest T., "Program Self-Distribution by Satellite: Some Key Practical and Legal Considerations," *Current Developments in Television and Radio,* The Practising Institute, New York, NY, 1982, pp. 289-312.

Kitty Morgan

6. Making The Connection: Distributing Original Programming

In the past few years the cable television industry has been acquiring a great deal of original programming made by independent producers. This programming ranges from local programs appropriate for public access channels, for which there is little or no license fee or recompense given for the material, to work suitable for the large national pay cable services, for which there is a license fee. The focus of this chapter will be on the kinds of programming cable television representatives purchase, and the options available to independent producers when they are trying to distribute original programming.

The telecommunications world is in an extremely volatile state in 1982. Cable systems are constantly changing, expanding, and merging. Several large cable companies are producing and co-producing their own series, specials and movies. Some experts predict that programming produced by the cable industry itself will eventually supersede acquisitions of existing and independently produced materials. As of this writing, however, there is a great need on the part of cable systems to acquire completed work of all kinds.

Kitty Morgan is the founder and President of Independent Cinema Artists and Producers (ICAP). She is also a free lance writer and producer.
Copyright 1983 by National Federation of Local Cable Producers (NFLCP).

The tremendous explosion of the number of satellites and satellite services is another indication of the volatile state of the telecommunications world. As of July, 1982, forty-seven program services are delivered by four satellites. At least fifteen more satellites are projected to be in orbit by 1985 and at least thirty more announced satellite services will be offered by 1985. Not all of these services will survive, but such numbers illustrate the potential outlets for distributing independently-produced programming.

This growth in distribution outlets makes innumerable the opportunities for marketing original programming. This article will concentrate on completed work available for leasing to cable systems. Material commissioned by a cable system or made with development money from the cable system is a different matter and will not be addressed here.

WHAT CABLE PROGRAMMERS WANT

In general, cable programmers prefer to acquire work that has fairly high production values and appeals to a national audience, even if it is a very specialized one. For example, a sporting event involving nationally-known teams would be preferred to the local high school baseball game. Local or regional programming (appropriate for a local public access channel) can and should be made, but it is not going to be of much interest to a programmer for a national cable service. Local cable channels provide both a valuable service for their own communities, and good experience for producers. Material for local channels can be made on a low budget; black and white films and one-half inch videotapes are more acceptable locally than nationally. National services usually accept only "broadcast quality" productions. Such broadcast quality productions are defined as those produced on sixteen or thirty-five millimeter film, three-quarter inch, one inch, or two inch videotape.

A wide range of programming is shown on the national services. Cable television has vastly increased the number of channels available to viewers since 1979. This abundance of channels has led to "narrowcasting" or programming for targeted or specialized audiences. It is now possible for all kinds of channels to be supported by specialized audiences, for example: channels for music, news, weather, sports, health-related material, religion, cultural, and arts programming. In addition, there are channels for minorities, children, and adults.

Shorts

The largest volume of independent production sales to television markets (referring to number of sales, not dollar amounts) is for shorts. Shorts are films and tapes less than one-half hour in length, and are very popular with most cable systems. They are used in the breaks between feature films, sports or specials, which are the main fare of most cable services. Most programmers refer to such material as "filler" or "interstitial" programming. Distributors prefer the phrase "complementary" programming.

Shorts ranging from animation through live action to documentaries are all of interest to cable services. Fast-moving, action-packed work with an obvious story line is most widely sought. Documentaries tend to be less marketable, as are extremely experimental works.

In general, a short is sold on a non-exclusive basis, which means it is possible to make a number of sales to different cable services, if the marketing is handled properly. Some cable services will request exclusive rights, for which they should pay accordingly. A few cable services will also insist that a program be a premiere or first showing. If the producer has already sold the program elsewhere, such a premiere sale, which could be more lucrative, may be lost. Usually this arrangement, called a premiere "window", is for an exclusive premiere and a certain number of repeats. In most cases, such arrangements are made only for features and specials, not shorts. In all cases, whether shorts, one-half hour programs, or features are involved, rates are determined by such factors as the size of the cable service, the number of repeats, the length of the license period, the kinds of rights being licensed, the type of material, and how strongly the programmer wants or needs the particular program.

Despite the popularity of shorts with cable services, many do not list the titles of shorts in their television guides. Programmers maintain that audiences are mainly looking for features, sports, concerts, and other specials, and don't consider shorts a draw. They also find it time consuming and costly to list all the shorts. Sometimes the affiliates, or the local cable systems that carry the programming services, distribute their own guides. These guides may include detailed shorts listings. Although it is true that few viewers subscribe to a service just to see the shorts, there are many who find the shorts the most innovative and stimulating material on a particular channel.

Cable services do receive letters about shorts, some of which are very enthusiastic. Supportive letters do have an impact. The more positive feedback the programmers get, the more likely they are to increase their budget allocations to license more shorts.

One-half hour programs are not as frequently purchased by pay cable as they are by basic cable or the networks. There are, however, some slots for half-hour children's shows, sports, health-related, cultural, and arts material, as well as a variety of specials and series.

Features

Independent features are also of interest to various television services. One should get as much "theatrical" exposure for such works as possible before trying to sell them to cable. Theatrical is used here to designate exhibition in theaters, not in the educational market, such as schools or libraries. The more festival awards and reviews the program has received, the more attractive it will be to program buyers, and the more likely to get a reasonable license fee. Features are of particular interest if there are recognizable actors, directors, and writers involved. Although independent features are up against popular Hollywood and foreign movies, the demand for programming is so great that even the lesser-known features have a chance. Particularly popular are children's and adult-only shows.

DISTRIBUTING YOUR PROGRAM

When distributing a program there are two options; letting a distributor sell it for you or selling it yourself. There are advantages and disadvantages to each option.

Most cable programmers prefer to deal with a distributor rather than an individual producer, although they will not close the door on anyone who has material they want. A number of factors make it easier for a programmer to deal with a distributor. For instance, a contract for several titles at once can be made with a distributor, as opposed to creating a separate contract for each individual producer's film or tape. This saves the programmer time and money. Also, distributors are already set up to deal with the contracts involved with the sale or lease of a program. Independent producers often don't understand such contracts so that the cable programmer has to spend a great deal of time explaining their stipulations. It is part of the distributor's job to take the time to explain the contracts to independent producers. The distributor is also familiar with a wide array of potential markets that would take anyone not in the business some time to learn in order to make the best possible choices.

Self-Distribution

An alternative to working with a distributor is distributing the program yourself. Producers who are intent on selling their own work should realize a large part of their time and energy will be spent on contact and negotiation work rather than in creating shows.

For those producers who are knowledgeable, there are a few advantages to self-distribution. Some producers like to know what is happening with their program at all times and feel they can better monitor sales if they do it themselves. Cable programmers will deal with an individual in such cases if the program is quite out of the ordinary. Sometimes a producer wants to put certain restrictions on the way a program is used (for example, that it not be used with commercial breaks). A producer may feel s/he can be the best advocate for such a position. Some distributors and cable services find such restrictions difficult. Other distributors, however, such as ICAP, have always been willing to consult with the producer about such matters.

There are also dangers in self-distribution. One who doesn't know how much to charge for a program may sell it for much less than it is worth at terms that are not in ones best interest. Not only is this a problem for that producer, it also undermines everyone else by driving the market price down. If in doubt about how to handle a sale, do not give away anything free; consult with a knowledgeable distributor or let someone else negotiate. Exposure is important, but being properly compensated is the key to keeping the market value of independent work as high as possible.

Before distributing your work (and preferably before production even starts) make sure all copyright laws have been observed. Do not use a Beatles' song unless you have a great deal of money to pay the record company or copyright owner for permission to use it. Keep accounts of all music used so that you can fill out a music cue sheet, which may be requested by the distributor or purchaser. The music cue sheet includes the title, composer, publisher, and length of the particular piece of music. If you have questions about the use of a particular piece of music it would be wise to consult with a lawyer or with a music licensing group such as BMI or ASCAP before using it.

Also make sure your own work is copyright protected. Do not exhibit the work or distribute it before putting the date, the word copyright, and the name of the copyright owner(s) at the end of the program and filing for the copyright with the Copyright Office. For more information about copyrighting consult a lawyer or the Copyright Office, Library of Congress, Washington, D.C. 20559.

CABLE PROGRAMMING SERVICES, NETWORKS, AND BUYERS

The following is a list of the major cable services that acquire independently produced material, and the types of work they are seeking. The list concentrates on entertainment-oriented outlets, rather than informational services, such as Cable News Network, which rarely use independent productions.

There are four general categories of services: national pay cable; regional pay cable; basic cable services, and subscription television or STV. The larger services of each category are listed first.

National Pay Cable Services

Home Box Office (HBO) is a Time, Inc. subsidiary with eight and one-half million subscribers. HBO is the oldest and largest pay cable service, consisting of all types of feature films, sports, and specials. It programs twenty-four hours a day, seven days a week. HBO uses many shorts of all kinds and all running times, although those less than twenty minutes are preferred. All genres are used if they are not too abstract or slow-moving. Fiction is chosen if the acting is good. Animation is accepted if it is not too experimental. Documentaries are used if they are fast-paced. Portraits of artists are not accepted unless they are very well known or unusual. The documentary division is separate from the shorts and intermissions department and does most of its own production. HBO is beginning to produce many of its own movies and series. HBO has minimal interest in independent features.

Showtime (a joint venture of ViaCom International and Group W Cable) has three million subscribers. Showtime programs twenty-four hours a day, seven days a week. It runs a variety of features, specials, Broadway productions, and concerts. Showtime does not use as many shorts as HBO, but is interested in comedy, music, and children's shorts. It emphasizes original series productions including adult soap operas and specials of all types.

The Movie Channel is a Warner Amex Satellite Entertainment Service with two million, two hundred thousand subscribers. The Movie

Channel programs only features and uses mostly Hollywood films, one foreign film a month, and a few independent features.

Cinemax, also owned by Time, Inc., is a separate division from HBO. Cinemax has one and one-half million subscribers. Cinemax programs twenty-four hours a day with a wide range of movies and shorts. Cinemax accepts more adventurous shorts than HBO.

Spotlight is owned by four MSOs (Multiple System Operators), including TCI, Cox, Storer, and Cablevision. Spotlight is distributed only to cable systems owned by these four companies, and in 1982 has just less than one million subscribers. A movie channel, Spotlight uses all types and lengths of shorts.

Escapade/Playboy is a joint venture of Playboy Productions and Rainbow Programming Service. With over two hundred thousand subscribers, it programs mainly R-rated features and is interested in compatible shorts of similar subject matter. Comedy shorts are also considered.

Home Theater Network (HTN) is principally owned by Westinghouse Broadcasting and has over one hundred sixty thousand subscribers. In 1982 this service will change from programming a few hours a day to twelve hours daily. At the point of this changeover, the service will be called HTN Plus. HTN is looking for family-oriented movies, as well as travel material. HTN wants and will buy G-rated programming of all lengths and types.

Galavision, owned by Spanish International Network, has one hundred thousand subscribers. It offers movies, specials, mini-series, and sports, as well as American movies dubbed in Spanish. All programming offered to Galavision should be of interest to a Spanish-speaking audience.

Bravo!, from Rainbow Programming Service (the same company that owns Escapade), has about fifty thousand subscribers. Bravo! was the first of the cultural and arts services created, and is the only one offered as a pay satellite service. In the past, Bravo! has acquired a number of artist's portraits and performance pieces. Bravo! is still interested in shorts dealing with the arts. Unlike other services that have an ongoing acquisition policy, Bravo! tends to program in spurts during specific buying seasons.

The Entertainment Channel, also referred to as RCTV, is a cultural service with exclusive rights to BBC material. In addition, the Entertainment Channel plans to use feature films, original dramas, children's programs, women's programs and shorts.

The Disney Channel will be a pay cable service of Walt Disney Productions, scheduled to be launched in April 1983. The Disney Channel will have children's and family programming, most of which will be produced by Disney Productions.

Regional and Mini-Pay Cable Programming Services

PRISM is a regional service out of Philadelphia and has over two hundred thousand subscribers. It is interested in shorts less than twenty minutes in length and has a preference for live action and animation rather than documentaries. PRISM also uses features, sports, comedy, and family-oriented programming.

Z Channel, a service of Theta Cable of California, wants shorts of all types to show between features.

VU-TV has twenty-five thousand subscribers in several states and needs shorts of all types. VU-TV prefers to do bulk buys, and has offices in Glendale, Arizona.

UPTOWN, a service of Telepromter, has fifteen thousand subscribers. UPTOWN has a small budget, but is interested in shorts and other independent programming.

Basic Cable Services

Getty Oil's **Entertainment and Sports Programming Network** (ESPN), reaches over fifteen million subscribers. ESPN is interested only in sports and sports-related programming.

USA Network is a Time, Inc., Paramount/MCA service reaching ten million subscribers. This is a general interest network with several specific programming areas, including: **Calliope**, a children's service which uses a variety of one-half hour and one hour formats; **Time Out Theater**, a sports-related service using one-half hour and one hour formats from independents; **Daytime**, composed of women's programming, most of which is not acquired from independents; **Night Flight**, a weekend collection of music and concert specials which uses some independent work; and **Alive and Well**, a health-related series. The **English Channel**, an educational and cultural service similar to public television, is looking for one-half hour and one hour documentaries from independents on a wide range of subjects including life styles, nature, and art. **Black Entertainment Television** (BET) is carried by USA Network seven days a week. BET programs features, with some music shows and documentaries of interest to black audiences. Occasionally independent work is used. USA Network also shows specials and a wide range of movies, many of which need shorts of all kinds and lengths.

Nickelodeon, a Warner Amex service, reaches seven million seven hundred thousand subscribers. A children's service, (although not directed towards preschoolers), the programs of Nickelodeon are often made by independents and include original drama, sports, Americana, and animation. Nickelodeon operates thirteen hours a day, and needs additional programming.

ARTS (Alpha Repertory Television Service), is a joint venture of Hearst Publications and ABC Video. ARTS reaches six and one-half million subscribers. ABC was the first network to go into cable, forming a cultural service using performing and visual arts material. Originally much of the programming was made in Europe, but ARTS is now interested in material from the United States. Independent work of high quality in the following areas may be considered: drama, music, ballet, opera, literary works, sculpture, painting, photography, and design. ARTS provides three hours of programming each night.

MTV (Music Television) is a Warner Amex service which reaches four million subscribers. MTV programs twenty-four hours a day with concerts, popular music programs, and record company clips. MTV's target audience is teenagers and young adults. Independent productions involving well known performers are of interest to MTV.

Subscription Television Service

ON-TV is programmed by Oak Media in California. ON-TV uses shorts of all kinds except documentaries.

SELEC-TV, a California based service, is also interested in independently produced shorts of all types.

EDITOR'S NOTE

Several agencies exist that sell or lease independently produced programming. Independent Cinema Artists and Producers (ICAP) is one such organization.

ICAP was created in 1975 specifically to distribute independent productions to cable. ICAP is a non-profit organization that works on behalf of independents and returns up to seventy-five percent of revenues to the producer. ICAP acts as an advocate for all types of independent work, including the more unusual material. This strategy has paid off. On several occasions ICAP has persuaded programmers to take offbeat films for which they received a tremendous amount of positive audience feedback.

Although ICAP has expanded to cover the full television spectrum, including networks, independent broadcast stations, public television, foreign television, and home video, cable continues to be ICAP's specialty. As of July, 1982, ICAP handles 800 titles of all types and running times.

If you would like ICAP to consider distributing your program, send a written description first, including: the name of the program, credits, format, length, release date, any promotional material, and a list of prior exposure of the work. Do not send any programs until ICAP has requested you to do so. For further information write: ICAP, 625 Broadway, New York, N.Y., 10012, or call 212-533-9180.

Another distribution agency is Coe Film Associates. Coe distributes programming to all forms of television, including the following: pay cable; basic cable; independent stations; educational and instructional markets; networks; and STV. Coe Associates sends programming to a number of countries, among them: Australia; Canada; England; New Zealand; and the United States. Interested in all subject matter of any length, Coe receives forty percent of the gross revenue for its work, and reports back to the producer with financial accounts every six months. Send a three-quarter inch videotape copy of your show for screening purposes, and allow two weeks screening time. For further information, contact: Beverly Freeman, Coe Film Associates, 65 East 96th Street, New York, NY, 10028, 212-831-5355.

K-PAY Entertainment, Inc. distributes primarily feature films and entertainment specials to all forms of pay TV including pay cable, STV, and MDS. On some occasions it does put up advances for production, and usually takes a percentage of the gross for payment. K-PAY has a one week turnaround time for screenings and prefers three-quarter inch videotape for screening. Contact Leonard Krane, President, K-

PAY Entertainment, Inc., 2049 Century Park East, Los Angeles, CA, 90067, 213-556-2633.
Film Gallery, Inc. distributes shows to pay TV markets in the United States. Film Gallery is interested in feature length programs of all kinds except documentaries. Film Gallery's payment for its services varies. On occasion it has put up front money to produce works. Film Gallery needs one week screening time and prefers three-quarter inch videotape. Write or call James Dudelson, Film Gallery, Inc., 500 5th Avenue, New York, NY, 10110, 212-944-2656.

Rental fees an independent producer may receive from a program will vary considerably. From twenty-five to two hundred dollars per minute is one estimate quoted, although much higher fees per minute, especially for features, may be obtained.

RESOURCES

Magazines

1. *Broadcasting*, Broadcasting Publications, Inc., 1735 DeSales St., N.W., Washington, D.C. 20036.
2. *CableVision*, Titsch Publishing, Inc., 1139 Delaware Plaza, P.O. Box 4305, Denver, CO. 80204.
3. *The Independent*, Association for Independent Film and Video Makers, 625 Broadway, New York, NY 10012.
4. *Multichannel News*, Unit of Fairchild Publications, a Division of Capital Cities Communications, Inc., 300 S. Jackson, Denver, CO.

Organizations

1. Association for Independent Film and Video Makers, 625 Broadway, New York, NY 10012, 212-473-3400.
2. Independent Feature Project, 80 E. 11th St., Suite 439, New York, NY 10003, 212-674-6655.

Publications

1. *Access II: The Independent Producer's Handbook of Satellite Communications*, National Endowment for the Arts, Media Arts Program, Mail Stop 552, 2401 E St., N.W., Washington, DC 20506, 1980. Copies available from The American Film Institute, John F. Kennedy Center for the Performing Arts, Washington, DC 20556.

2. Mahoney, Sheila, DeMartino, Nick, and Stengal, Robert, *Keeping PACE With the New Television*, The Carnegie Corporation of New York, 437 Madison Avenue, New York, NY 10012, 1980. Currently out of print.

Margie Nicholson

7. Cable Television Advertising

Cable television in the 1980s is a service business—service to the customer, service to the community, and service to local businesses through the development of advertising sales. Cable can be a new and exciting advertising tool for local businesses; it can also be a new and exciting way to increase revenues for your cable system.

Cable advertising can help a local business generate awareness, promote a specific product or service, and target messages to a specific geographic and/or demographic audience. Local businesses are looking for new ways to reach consumers with their promotional messages; you can help by showing them how to creatively and effectively use cable advertising.

Although the cable industry still has much to learn about advertising sales, many systems have invested the expense and effort necessary to start their advertising operations, and are looking at initial annual revenues of three dollars to ten dollars per subscriber. Establishing advertising sales in your local system will require research and planning

---

Ms. Nicholson is currently the Coordinator of System Development for U.S. Cable of Northern Indiana. Prior to this post Ms. Nicholson managed a full-service advertising agency called The Agency; wrote and produced commercials; and worked as Community Program Director for Viking CATV Associates in Wisconsin.

Copyright 1983 by National Federation of Local Cable Programmers (NFLCP).

for the best approaches and methods for your situation; taking an inventory of your channels for potential sales outlets; setting up internal systems for monitoring and controlling the advertising operation; hiring and guiding sales personnel; and promotion of your advertising services.

RESEARCH AND PLANNING

Look before you leap into cable advertising sales. You'll want to research your local economy, businesses and potential advertisers, competition for advertising dollars, and subscribers. Make contacts within the industry and through the Cabletelevision Advertising Bureau to find out about the turmoils and triumphs of other system operators. This initial research will help you develop a financial plan with projected revenues and expenses.

Get to know your local economy. What is the present and predicted condition of the local economy? How's the employment rate? What companies or industries are the primary employers and what is their economic outlook? Do residents patronize your local businesses or go elsewhere for their goods and services? Is business generated from within your system area or drawn from a region outside your subscriber base?

Find out about your potential advertisers by studying what they are doing in the other local media. Clip or take notes on newspapers, radio, television, billboard, direct mail, and other advertising vehicles. Hang on to any coupon books, shoppers, and other ad-supported materials that cross your desk—including program guides for local theatre and sports events. Use these materials to generate your prospects list and estimate the potential for advertising revenues on your system.

Note the size of each ad, the frequency, the type of promotion used, and the environment chosen by local businesses for their advertising messages. Be a consumer in your community. Visit local stores, try their products and services, talk to them about their advertising budgets and results. What are their problems? "Let your fingers do the walking," through the Yellow Pages to determine which businesses choose to be listed, and the type of promotional messages they employ. Studying the Yellow Pages will also give you an idea of which advertisers are using "co-op" dollars; this will be important as you start selling. If the advertiser exhibits the use of a manufacturer's product in the ad, this is a good indication that co-op dollars are being used. Co-op dollars is money set aside by a manufacturer that retailers can use to advertise their products. Sometimes the payment is in the form of money and sometimes it is in product form.

Let your competition try to sell to you. Call and ask them to send over a sales representative with information on ratings, audience demographics, and rate cards. Question them about the types of promotion they are using, and which programs and program formats are hot. If you are unfamiliar with ratings, shares, CPM (Cost Per Thousand), and have never confronted the endless columns of Arbitron or Nielsen ratings before, let them teach you how to compute and understand these figures.

Do the local media representatives sell "by the book"? Do they stick to the rate card or negotiate? How much does it cost to produce an ad? What do they have to say about their competitors? Ask for any and all surveys, circulation numbers, lists of advertisers, and promotional literature. If you're thinking of hiring an account executive from among the local talent, take a good look at the person who's trying to sell to you.

Pay special attention to the newspaper circulation in your area. Newspaper coverage of the market has been dropping steadily since 1970; you may find that your cable system has greater penetration than the local paper.

Make sure you know your own system—how many households are passed, the number of basic subscribers, the percentage of total passings penetrated, and the projected growth. Find out how other cable companies have developed and used their surveys. Gather research and demographics about the cable subscriber from national trade publications. Conduct a local survey or ascertainment interviews with local leaders that can be used as ammunition for your cable sale. Next, conduct a mail survey, phone survey, or focus group to find out more about your subscribers and what they think about your system and programming. Cable subscribers are a quality audience with disposable income (they're already purchasing your cable service). Potential advertisers may want to know about their buying habits; whether they own or rent, their family size, their travel habits, and about their credit cards. The more information you have about your audience, the better prepared you will be when some media hot shot starts firing those difficult questions.

Take advantage of the experimentation that has already been done in many cable systems across the country by making contacts within the industry and talking to them about developing your advertising sales strategy. The Cabletelevision Advertising Bureau can provide a wealth of information from system profiles and sample forms to rate card strategies and sales tools. Cable trade publications will frequently profile successful advertising sales efforts. You may want to call those systems directly, particularly the ones with characteristics similar to yours, and ask them to send you their rate cards and other sales promotion literature. Definitely contact any other cable systems in the area who are selling advertising; you may be able to help each other with leads and strategies.

Part of your planning should include a financial analysis with projected income and expenses. Your plan should also include projected costs for equipment, personnel, office supplies and overhead, promotion, and marketing. Don't forget to include projected agency commissions and bad debts. Determine what your production costs will be and how they will be billed. If you have existing production facilities, you'll probably want to produce commercials in-house. If you don't have existing production facilities, you may want to sub-contract with a local production house, independent producer, or broadcast television station. Providing well produced, low cost ads to advertisers will be a good incentive for businesses to purchase time.

The following is a simplified first year financial planning sheet for a cable system with twenty thousand subscribers, where all production is billed at cost to the advertiser (production costs are not included). This chart is a hypothetical one that illustrates how a cable operator can begin to project advertising revenues and what starting costs might include. The average advertising revenue that could be expected per subscriber in any given system in 1980 was one dollar and seventy-three cents. This figure was based on industry averages and projections for systems first researching and initiating such advertising programs. This chart starts with a more realistic (for 1982) three dollar figure per subscriber, and illustrates the results of six and ten dollars per subscriber. Any of these figures may be reasonable projections for your system, depending on your marketing situation (such as how many television and radio stations exist in your area) and your management plan. The subscriber revenue figure selected (i.e., three dollars) then becomes a measure of your advertising program effectiveness and how realistic your goal is (sixty thousand dollars in gross revenue), and your assessment of your area's marketing possibilities. Following this example, a forty-eight thousand dollar loss would be incurred in the first year of operation, which could turn into a profit in the fourth year (by steady growth in subscriber revenue figures and through the amortization of equipment). By doing the initial research described in this section, and modifying this sheet with your needs and figures accordingly, you can reach your own conclusions about the possibilities of subscriber revenue and expenses in your situation.

FINANCIAL PLANNING SHEET, CABLE SYSTEM WITH 20,000 SUBSCRIBERS

Projected Revenues
Anticipated or Projected Revenue

Per Subscriber	$ 3	$ 6	$ 10
Setting Sales Goals:			
$ Ad Sales per day	$ 164	$ 329	$ 548
# 60 second spots per day at $20 each	8.2	16.4	27.4
OR			
# 30 second spots per day at $15 each	10.9	21.9	36.5
Gross Revenues	$ 60,000	$ 120,000	$200,000
Projected Expenses			
Sales Expense (estimated 40%)	$ 24,000	$ 48,000	$ 80,000
Sales Support (traffic, billing, secretarial)	$ 4,000	$ 4,000	$ 4,000
Playback (6 part timers)	$ 42,000	$ 42,000	$ 42,000
Insertion Equipment (manual for 3 channels amortized over 3 years)	$ 23,300	$ 23,300	$ 23,300
Office/Phone/Overhead	$ 4,800	$ 4,800	$ 4,800
Packets, Survey, Launch	$ 4,800	$ 4,800	$ 4,800
Agency Commissions (estimated 15% of 30% of sales)	$ 4,500	$ 6,000	$ 10,000
Bad Debt (1%)	$ 600	$ 1,200	$ 2,000
TOTAL	$ 108,000	$ 134,000	$170,900
NET	$(48,000)	$(14,100)	$ 29,100

Finally, you should extend your plan over a five year period to incorporate system growth, payback on capital expenses, and additional personnel. Use your research and financial planning to set the necessary, realistic sales goals to attain your desired results.

INVENTORY YOUR CHANNELS FOR POTENTIAL SALES

Take an inventory of your current channel line-up and all of your resources to see where ads can be sold. You'll want to consider the automated channels, local origination channel(s), and satellite channel availabilities. Are you interested in developing and selling "infomercials"? If so, on what channel can they be run? What about a program guide? A monthly guide can be an attractive advertising medium and an opportunity to generate some additional ad sales revenues.

Automated Channels

Selling listings on your automated channels can be an easy first step into the world of cable television advertising. Several national data channel services will permit you to insert full page or crawl advertising into their programming. If you are already programming your own automated channels, the addition of advertising should mean very little additional time and effort. Keep in mind that one-third of total advertising revenue to newspapers comes from classified advertising; this should be a good incentive to turn those channels into money-makers.

Make up a form with blank spaces so that customers can fill in their own messages. The forms, along with pre-payment, can easily be turned into the front office and then forwarded to the traffic coordinator for insertion onto the channels. Keep the rates simple (i.e., three lines for three days for three dollars) and in line with local classified advertising rates. Check your equipment in advance to ascertain how many pages or crawl messages can be inserted and let the office personnel monitor the inventory as they sell.

Some examples of automated channels include:

Time/Weather. This channel usually has a high cumulative viewership and, if it is programmed locally, there is a great deal of inventory to sell.

AP or UPI News. You may want to insert local news announcements as well as commercial messages to make this channel of particular interest to your local viewers.

Cable Guide. This channel is frequently checked by cable subscribers for up-to-date programming information. Insert ads and promotions for your local programs.

Community Calendar. Local community groups and organizations will send you press releases to use in programming this channel. You may want to develop games, cable trivia, or sales promotions with local merchants for this outlet. On a cable trivia or quiz show subscribers call in with answers and you can reward them with giveaways, coupons or prizes provided by local merchants. This is good promotion for local merchants and allows you to build viewership, reward viewers, and get some idea of the audience response to your programming.

Financial Channel. Providing New York and American Stock Exchange information on a delayed basis will attract an upscale viewer that banks, insurance businesses, real estate agencies, investment advisors, brokers, and other advertisers will want to reach.

Depending on the number of available channels on your system, and the size of your market, you may want to develop additional specialized automated channels targeted to specific audiences, such as real estate listings, travel and entertainment information, and shop-and-swap channels.

Local Origination Channels

If you have an existing program schedule on your own commercial channel, you will definitely want to sell advertising to help underwrite the costs of production and your Local Origination (L.O.) department operations. Programming that franchise agreements require you to provide could be partially or fully underwritten by local ad sales (i.e., programming such as news or sports). There are several alternative methods for developing advertisement-supported L.O. programs.

One method is to produce your programs in-house by paying all talent and production fees and then selling full sponsorship or advertisting availabilities. A weekly sports series or "Game of the Week" could be produced and sold in this manner. Or your local Chamber of Commerce might want to produce a weekly business show. It would provide the host and guests while Chamber business members would buy advertising to support production and cablecasting. A local arts association may ask its business benefactors to sponsor a weekly arts promotion program. You may discover a local radio talk show host who wants to branch out into cable and bring advertisers along. Establish your program plans far enough in advance so your account executive (A.E.) can pretest ad sales interest. You may want to hold a special screening to show potential advertisers what programming looks like.

A second alternative method to developing advertisement-supported L.O. programs is to produce shows in cooperation with individuals, businesses or local institutions. You may cover the costs of production while they would serve as producer or talent, and ad revenues could be split. Once you announce your plans for local program production, you are likely to be faced with a deluge of people who want to make their own television programs, or more specifically, want to be television stars. You'll have to judge their abilities very carefully to determine which will be able to attract a local audience and advertising support.

A third method is an option for those cable programmers with existing program inventory or acquisition budgets. A cable programmer in this position may want to run, rent, and/or purchase programming that could be advertiser-supported. Sound out potential advertisers first to see if they'll make a commitment to sponsor the program or series. Try creative packaging by linking a program with a sponsor (i.e., a sports series with sports equipment stores and manufacturers) or grouping programs for targeted audiences that will be attractive to advertisers. Apparently Fred Silverman got his start at WGN in Chicago by putting a bunch of dusty titles in the film library together with a genial host and calling the series, "Film Classics." It was a great success. So think about digging through your archives, researching tape loans or rentals, or cooperating with other cable systems to put together a creative package. If it worked for Fred . . .

Making use of barter programs is a fourth method of achieving advertisement-supported L.O. shows. For example, a money management series was offered to Chicago area cable operators on a barter basis. Cable operators got a financial series and a thirty-second spot to sell; the producer developed and distributed the series and sold the other advertisements. One cable operator, U.S. Cable, sold an exclusive sponsorship to a local insurance agency and made an easy four hundred dollar profit. If barter programs are available to you, insist on getting local advertising availabilities to sell. Why give away your valuable channel time and let someone else collect the revenues? If you're very ambitious you might want to produce your own programs, distribute them to other cable operators in your area, and sell the commercial time to advertisers who want to reach that regional market.

Infomercials and Cable Shopping Guides

One major advantage of cable television is that there is no real limit on the length of advertising messages. This provides advertisers and the cable operator with the opportunity to create program-length commercial messages, otherwise known as "infomercials." Each advertiser could create its own infomercial or the cable operator could develop special shoppers' programs by grouping commercial messages together.

Village Cable in Chapel Hill, North Carolina, runs a Home Shopping Channel from six to eleven each evening with infomercials from local businesses. According to President Jim Heavner, advertisers must run their infomercial a minimum of ten times per month at a charge of twenty-five dollars per one-half hour. Production charges are extra and usually run between two hundred dollars and four hundred dollars for a one-half hour program. Infomercial subjects have included real estate, retail clothing, solar heating, stereos, computers, and travel. Often local businesses, like stereo dealers, will be able to get film or slide presentations from a vendor or manufacturer. These can be transferred to videotape with an introduction and a conclusion by the local dealer to create an effective, low cost infomercial.

The cable operator might want to group infomercials around a special theme, similar to a newspaper's "Home" or "Wedding" sections. Production Manager Steve Craddock notes that Village Cable's best shows have been the Spring Fashion Show featuring three different stores, and the Christmas Show in which ten stores displayed their Christmas specials.

The Film Workshop Council at Manhattan Cable produces another variation of cable shopper's programming. The Council produces one minute interviews with local shop owners. The resulting commercials run twice each night, seven days a week, for six weeks as part of the Cable Shopper's Guide. Total cost to the advertiser for production and channel time is six hundred dollars.

Satellite Services

Many satellite programmers offer local advertising availabilities and encourage local systems to sell advertisements on their channels. From one to seven minutes per hour of advertising availabilities are provided by program services like BET, CNN, CNN2, ESPN, MSN, MTV, SIN, SNCI, SNCII, SPN, and USA. Contact them for information, promotional materials, and any audience research that you can use to sell local businesses on the idea of sponsorship. Calculate the number of minutes which can be sold on each channel and the approximate value of those minutes to local advertisers to determine which channels to sell. You'll want to get the highest possible return on your initial investment in commercial insertion equipment.

A major decision will be whether to go with manual or automated commercial insertion equipment. Current opinion seems to be divided. Some operators are willing to make the initial extra investment for automated equipment which can cost over sixty thousand dollars per channel. Others prefer the control of inserting commercials manually, assuming that this will mean less frequent errors. According to Mr. Randy Van Dalsen, Community Program Director for United Cable Television, his company will use manual insertion until tone cues are standardized and there are fewer resulting errors. Mr. Van Dalsen also noted that use of the automated insertion equipment requires extra pre-production time. Manual insertion equipment costs between twenty thousand and thirty thousand dollars per channel in addition to the extra personnel expenses.

Basic and premium program suppliers may also be a source of local advertising revenue. Many such suppliers offer co-op dollars and launch support funds for local advertising which could be placed on your channels or in your cable guide. U.S. Cable's National Marketing Director, Ron Russo, emphasizes that it is important to review your contracts with program suppliers and negotiate the use of promotional dollars in your system or guide.

Program Guide

A monthly program guide is an excellent way to reinforce subscriber satisfaction with cable service, promote local programming, and generate revenues through local advertising sales. The cable guide stays in the home and is used, probably on a daily basis, for an entire month. Often cable viewers have nowhere else to look for cable television program listings.

Your cable guide is a promotional vehicle for your company, your L.O. department, and your sales department. Businesses who are unfamiliar with cable and are reluctant to try the new electronic media, may be introduced to cable via advertising in the guide. Oftentimes businesses have advertisement slicks ready to go when your salesperson walks in the door. These can be put directly into your guide as coupons, sales promotions or advertisements. Surveys, sweepstakes, contests, and coupons included in the guide, whether they're run by you or your local businesses, will generate subscriber interest and help you measure and understand your potential audience.

Set your cable guide rates so that they are competitive with the cost of advertising in the local paper or television listings magazine. Compare their circulation and service with yours and set your rates accordingly. Remember to set a higher rate for the back page as it is a location of high appeal to advertisers. If you have someone with graphic design experience on your staff let them design the ads; if not, subcontract to a local commercial artist, community newspaper, or shopper.

SYSTEMS OF MONITORING AND CONTROL

Before you go out to sell that first advertisement you'll need to set up policies and procedures for monitoring inventory, sales, traffic, billing, scheduling, cablecasting, and collections. The following is a step-by-step scenario of how a system of monitoring operation works. In some cable company operations several of the following positions described may be fulfilled by the same person.

General

All Account Executive (A.E.) contacts need to be recorded on a daily log. The A.E. then submits compiled logs to the Ad Sales Manager each Monday which reflect the preceding week's work. The Ad Sales Manager maintains an updated inventory of availabilities and submits a Monthly Sales Report which reflects the total sales contacts made, the number of availabilities, and the percentage of availabilities sold.

Sales/Traffic

The A.E. is responsible for completing contract confirmation with the client. The client indicates acceptance of the agreement by signing it. A copy of the agreement is sent to the client for his/her records.

The Ad Sales Manager reviews, approves, and forwards the contract confirmation to the person responsible for Accounts Receivable. The contract is given an account number and an account ledger card is initiated.

Accounts Receivable retains one copy of the contract for the account file and forwards one copy to the A.E., the Ad Sales Manager, and the Traffic Manager. On the following page see the illustration of a contract used by U S Cable.

The Traffic Manager prepares a cablecast log, assigns a tape number, and verifies that the commercial is received and/or prepared.

The person cablecasting the commercial initials a log verifying the running of the contracted material and forwards the completed log to the Traffic Manager. If for any reason an advertiser's spot is not properly cablecast, there should be every effort made to run that spot as soon as possible, and when the advertiser agrees it should be run. These are called "make-goods." These make-goods should also be noted on the log.

The Traffic Manager verifies this log against the contract confirmation and forwards the log to Accounts Receivable.

CONTRACT

US Cable of Northern Indiana
Advertising Services Division
821 W. Glen Park Avenue
Griffith, Indiana 46319

Account No. _____
Invoice No. _____

Account Exec.	Order No.	Date	Page

Advertiser _____ Contact _____
Agency _____ Product _____
Invoice to _____ Co-op _____
Address _____ Production Source _____
City/State/Zip _____
Start date _____ Stop date _____

Start	End	Quantity	Length	Rate	Description/Rate Plan	Totals

Total _____
% Commission _____
Total Due _____

Time	M	T	W	TH	F	S	SU	Special Instructions

Accepted for US CABLE _____   Accepted for CLIENT _____

WHITE - Accounts Receivable; GREEN - Traffic; YELLOW - Account Executive; PINK - Advertising Manager; GOLD - Client

Invoicing

Accounts Receivable then prepares an invoice from the data on the log, and any existing ledger cards with a previous balance. Past due balances should be designated with a thirty, sixty, ninety, one hundred twenty, or one hundred thirty days notation.

Cable TV Advertising / 77

Accounts Receivable follows through by issuing invoices on the tenth and twenty-third of each month, retaining one copy for the account file, and forwarding remaining copies to the client, A.E., Ad Sales Manager, and Corporate Accounting. The following is an example of an invoice form.

PLEASE PAY FROM THIS INVOICE
MAKE CHECKS PAYABLE TO:

US Cable
of Northern Indiana
Advertising Services Division
821 W. Glen Park Avenue
Griffith, Indiana 46319

Invoice No. _N⁰ 0039_
Account No. _____

Account Exec.	Order No.	Date	Page

In Account With: _____   Advertiser: _____

Date	Day	Time	Length	Description	Quantity	Rate	Total

I hereby certify that the announcements specified on this invoice were Cablecast/Run on the dates and the times shown.

Subscribed and sworn before me this _____ day of _____, 19____.

Notary Public _____
My commission expires _____   General Manager _____

Total: _____
Total Due: _____
Date Due: _____

WHITE - Client; GREEN - Accounts Receivable; YELLOW - Account Executive; PINK - Advertising Manager; GOLD - Corporate Account

Collection

All payments are processed by the System Cashier. Once the Cashier receives a payment, one copy of the receipt is forwarded to Accounts Receivable for a ledger card update. All monies received are deposited in the system's Advertising Account. All monies received by the A.E. are reflected on the contract confirmation. Accounts Receivable then updates the ledger from the contract confirmation.

All receivables should be aged from the date of the invoice. A standard set of collection procedures can be employed at regular intervals beyond this date. At the thirty day mark, a second notice copy of the invoice should be sent to the client, the A.E., and the Ad Sales Manager. After sixty days a past due notice should be sent to the client, the A.E., and the Ad Sales Manager. At this point it is within the Ad Sales Manager's discretion whether to allow any further advertising by this client. The A.E. should make personal contact with the client to collect the past due balance. If the collection is not made within ninety days, the A.E. loses his/her commission. Another past due notice is sent to the client, the A.E., and the Ad Sales Manager after ninety days, but now Accounts Receivable attempts to collect on the overdue bill. After one hundred twenty days a notice is sent to the Ad Sales Manager, and Accounts Receivable, which makes a final attempt at collection. Unless there is a specific authorization written by the Ad Sales Manager and the General Manager, the account will be written off at this time. Unless otherwise instructed by the General Manager, the overdue account will be sent to a collection agency after one hundred thirty-five days.

Reporting

A Monthly Sales Report compiled by the Ad Sales Manager should be submitted to the General Manager and Corporate Operations no later than the thirtieth of each month. Here is a sample Monthly Sales Report Form.

Cable
of Northern Indiana

ADVERTISING SALES REPORT

_____, 198__

Sales Category	Sold		Invoiced	
	Current Month	Y.T.D.	Current Month	Y.T.D.
Spot				
Direct				
Agency				
Sub-Total				
Classified				
Direct				
Agency				
Sub-Total				
Cable Guide				
Direct				
Agency				
Sub-Total				
Production				
Direct				
Agency				
Sub-Total				
Program/Lease				
Direct				
Agency				
Sub-Total				
Other (List)				
Sub-Total				
TOTAL				
Actual				
Budget				
Variance				

821 West Glen Park • P.O. Box 69 • Griffith, IN 46319 • (219) 924-5005

A Monthly Accounting Status Report, which is compiled by Accounts Receivable, should be submitted to Corporate Accounting immediately following the twenty-third invoicing procedure.

A Monthly Aged Money Report compiled by Accounts Receivable should be submitted to Corporate Operations with the Monthly Accounting Report. The following is a depiction of an Aged Money Report form.

US Cable
of Northern Indiana

ADVERTISING AGED MONEY REPORT

_____, 198__

Client	0-30 Days	31-60 Days	$ Aging 61-90 Days	91+ Days	Written Off
TOTAL					

821 West Glen Park • P.O. Box 69 • Griffith, IN 46319 • (219) 924-5005

HIRING AND GUIDING SALES PERSONNEL

There is some controversy about how to select an A.E. for local cable sales. On the one hand, people with experience in radio or other media sales are thought to be a good choice. Presumably, they know media, they know the market, and they know the accounts and the needs of those accounts.

On the other hand, personnel evaluation experts like Jean and Herbert Greenberg suggest that there are criteria more important than experience in selecting sales personnel. In a February 1, 1982 "TVC" article, they indicate that empathy and ego drive are requisite qualities for all sales people. According to the Greenbergs, an excitement about the newness of cable and the ego strength to take rejection would be additional important qualities for successful cable salespeople.

Whether you raid your media competition or decide to develop new talent, your new A.E. will need an introduction to cable—and guidance from you before starting. It will be helpful for him/her to take a look at the local and satellite programming which will be sold and to spend some time with the production crew to learn how commercials are put together. The salesperson will need business cards, a phone, and some time to review the information gathered in your initial market research. Get together a prospects list and ask all your personnel for sales leads.

Telephoning such leads and "cold calls" can be used in developing prospects and approaching local businesses. Prepare a packet of information about the cable system, market area, cable subscribers, programming, and advertising opportunities. Providing information in a packet will make it easy to target promotional literature to each business and change or update the material as the need arises.

The sales presentation should be rehearsed and the approach to each new business should be well thought out. The A.E. should be aware of the following cable advertising sales pitfalls:

Don't be inflexible. Don't give the same sales pitch to each client. Listen to the clients' questions and let them tell you what they need to know to make a decision. If they like radio, talk about putting pictures with their words. If they buy billboards, tell them about your automated channels. If they're sold on broadcast television, talk about target audiences and upscale viewers.

Don't ask for something they can't give. Don't ask for all of the client's advertising dollars. Don't ask them to run ads in the wrong environment (i.e., kitchen appliances on your sports program), or during the wrong season. Don't push for a buy for next week if their media budgets were finalized six months ago.

Don't alienate the advertising agencies. Offer a commission to advertising agencies. Call on them and send them mailings and clippings. An advertising agency that is educated and enthusiastic about cable can bring you several new clients. Unless it's absolutely necessary, don't go around the agency to get to the client. The agency will never forgive you and if you ever do make a mistake in serving that account, the agency will make sure the client never forgets it.

Don't promise what you can't deliver. Don't oversell the benefits of cable. Don't persuade the client that a thousand widgets will walk out of the store the day after the ad runs. If those widgets don't move like Marines, you can say goodbye to your next sale. Keep your eye on that future business and build trust.

Don't get caught up in the numbers game. This is called "selling by the book," and you'll want to avoid it as much as possible. Naturally you'll want to present any favorable surveys or audience statistics to potential clients. But at this point in cable's development it probably won't be easy to sell by the book. Think of all those little radio stations in big markets who thrive even though their ratings are eclipsed by the giants. They target, negotiate, develop creative packages, and work for those sales. So should you.

The emergence of cable as an advertising medium has thrown discussions of ratings and methodologies up for grabs. Prevalent methodologies do not incorporate the demographics, psychographics, and sales measurements that will be necessary to get a true picture of advertising potential or effectiveness on a multi-channel cable system. Until cable audiences become larger in number and more sophisticated measurements are developed, sell what you have. Sell your upscale subscribers, targeted geographic area, environment, and the "magic of television." Sell your circulation, like the newspapers. Sell frequency and saturation. Sell cable as a complementary media buy. Sell the concept of cable as a new media. Sell experimentation. Finally, sell the "get there first" idea.

PROMOTING YOUR ADVERTISING SERVICES

Promote your advertising services by holding an advertising sales "launch" and an annual advertising party. Obtain distributing and cablecasting testimonials from successful advertisers, and develop a creative look for your local advertising messages.

The purpose of the launch is to introduce potential advertisers to cable television and show them your facilities and services. Promote the launch with several direct mail pieces to potential advertisers and follow up with phone calls to key members of the business community. Your launch should generate excitement and interest in the community and, above all, you must give advertisers an incentive to buy right away. You can do this by offering a unique advertising package, a limited time offer, a discount on production costs, and/or any other special inducements.

In an issue of *Cable Management Report,* the Cabletelevision Advertising Bureau profiled a very successful sales launch conducted by Village Cable in Chapel Hill, North Carolina. This forty-three hundred subscriber system generated one hundred fifty thousand dollars in advertising commitments from a luncheon presentation in July, 1981. Village Cable offered a choice of two Charter Cable Advertiser Plans "to the first fifty advertisers who wish(ed) to participate or all who initial(ed) at today's (that day's) clinic." Plan A, at five hundred dollars per month, consisted of a minimum of fifty thirty second spots or thirty-five sixty second spots plus five hundred dollars credit for production costs and a rate guarantee for one year on the first five hundred dollars each month. Plan B, at three hundred dollars per month, included a minimum of thirty thirty second spots or twenty-two sixty second spots with a rate guarantee for one year on the first three hundred dollars each month. Both plans were annual agreements for twelve consecutive months, cancellable after ninety days with no penalty.

The advertising sales launch and annual advertising parties should include videotape viewing of segments of your local programming along with clips from the satellite program services you will be offering to advertisers. Invite representatives from the satellite channels and your local on-air talent to attend and speak. Give potential advertisers all the information and reassurance they need to make an immediate buy. After the launch, follow up on those who were unsure with individual or small group sales presentations.

Monitor the progress of businesses who are using your advertising services and promote their successes. Include their success stories in your program guide. Produce special testimonials about the effectiveness of cable advertising to use on your channels and for sales presentations. Nothing sells advertising like results. Be sure to promote yours.

Finally, creativity will be an important part of your work in producing local commercials, developing sales strategies, and establishing an advertising sales department. Unlike our broadcast colleagues, most entrepreneurs in cable advertising will have to work with smaller production budgets. In addition, there is a much more challenging environment to sell in with so many competitors for the advertising dollars. This is where your creativity and willingness to experiment will be important. Your pioneering efforts to provide cable subscribers with creative, local commercial messages, and provide the business community with cost-effective advertising opportunities, may help our industry better understand and develop this new revenue source and business service called cable television advertising.

RESOURCES

Publications

1. *Cable Age,* 1270 Avenue of the Americas, New York, New York 10020.
2. *Cable Marketing,* 352 Park Avenue South, New York, New York 10010.
3. *Cable TV Advertising,* Paul Kagan Associates, 26386 Carmel Rancho Lane, Carmel, California 93923.
4. *CableVision,* Titsch Publishing, Inc., P.O. Box 5400 T.A., Denver, Colorado 80217.
5. *TVC,* Special Advertising Issue, Feb. 1, 1982, Cardiff Publishing, Circulation Service Center, P.O. Box 6229, Duluth, Minnesota 55806.

Associations

1. Cabletelevision Advertising Bureau, Inc., 767 Third Avenue, New York, New York 10017; (212) 751-7770.
2. National Cable Television Association, 1982 Cable Advertising Directory, Media Services, 1724 Massachusetts Avenue, N.W., Washington, D.C. 20036; (202) 775-3611.

Article

1. Alter, Robert H., "Cable Advertising" chapter in Hollowell, Mary Louise (ed.) *The Cable Broadband Communications Book, Volume 3, 1982-1983,* Communications Press, Inc., 1982, 1346 Connecticut Avenue, N.W., Washington, D.C. 20036.

Richard Wheelwright and Wm. Drew Shaffer

8. How the Cable System In Iowa City Works—Toward An Interactive Future

Mr. George Stoney, Co-Director of the Alternate Media Center at New York University (N.Y.U.), is considered by many to be the Founding Father of Public Access cable television programming. Mr. Stoney tells how his first experiments in two-way interactive cable were made possible in Reading, Pennsylvania, by a National Science Foundation (N.S.F.) grant and with the cooperation of the local cable operator, Mr. Earl Haydt, of BerksCable Company (who is currently Community Relations and Public Affairs Director with the American Television and Communications Corporation or ATC) who supported the concept of public access and who found that it interested people in cable. The 1976 N.S.F. grant, given to the City of Reading, the Alternate Media Center at N.Y.U., BerksCable Company, and a senior citizen organization, enabled them to cooperatively plan, design, and build an interactive cable system. Every home would have two wires instead of one.

Mr. Richard Wheelwright is Professor of the Media Arts Department of the University of South Carolina in Columbia. Mr. Wheelwright has been a visual communications consultant for over ten years as well as a professional writer.
Mr. Drew Shaffer is the Broadband Telecommunications Specialist for the City of Iowa City, Iowa. Mr. Shaffer has worked as a producer and consultant for ten years, and has taught at the University of Iowa.
Copyright 1982 by National Federation of Local Cable Programmers (NFLCP).

One wire could bring signals into the home from the station. The other wire could conduct signals from the home back to the station. Two-way television. People could participate in programs from their homes.

Ms. Red Burns and Mr. George Stoney, Co-Directors of the Alternate Media Center, worked on the experiment with all the other local participants to embark on a new vision of the future: if people were attracted to cable because of the community-produced programs, how much more would they be attracted if they could actually participate in programs and if cable could offer other two-way services? What forms might this service take? Who might benefit from these services?

To a large degree, the experiments in two-way interactive programming were made possible by the cooperation of the cable operator. All sides profited. It takes many different kinds of people to make community programming work day-to-day, they must work cooperatively, and they must find satisfaction . . . and for those people to shape and perform experiments in new methods of communication, they must share a common vision, although they may pursue it from different perspectives. Other cities have found this to be true too.

GETTING WIRED IN IOWA CITY

For example, in the heart of America, in Iowa City, Iowa, the citizens have designed a cable system that technically and administratively functions well for community programming. They have developed community video training to produce that programming, and they have experimented to discover methods of communicating more effectively within the community and to interact with the world outside the city limits.

To create a system that would function this way, the citizenry obliged their government to find a cable company that would agree to operate to some extent to the benefit of the community. This would include maintaining public access production facilities. The city government, with the aid of the Cable Television Information Center (CTIC) and the National Federation of Local Cable Programmers (NFLCP), devised an effective cable ordinance to insure that the company would adhere to the agreement. A city cable commission (the Broadband Telecommunications Commission or BTC) was formed to see that the cable ordinance is enforced and to insure the franchise fees the city received from the cable company are to be used for access facilitation, promotion, and company regulation.

Holding public hearings, creating the necessary study groups, determining the best ownership form, and community input were involved in laying the foundation so the best possible situation could occur. A great deal of political, administrative, and technical expertise was invested.

Several questions were posed to Iowa Citians: where do we put the public access and local origination production centers so that people can use them? What type of production equipment do we buy, considering that it has to be simple to use, yet give relatively high cablecast-quality performance? Who will be responsible for the public access production centers, and encourage and train people who want to be involved in community programming? Who is formulating and presenting such questions to the community and putting the decisions into action? The answer to that question is anyone who takes an active interest—the grassroots, the *vox populi,* finding new means to strengthen and benefit itself.

Iowa Citians realized they were forming a nervous system for their city. A nervous system should be controlled by a brain. So the public access and local origination centers were placed in the public library, easily accessible in its downtown location. People learned how to use video equipment in library classes (conducted by the city staff), and/or cable company classes, and then could check the equipment out like library books (from two centers in the library; one operated by the company, the other by the library).

Three different groups with three different sets of interests and expertise have been brought together—the cable company, the city government, and the library—to make community programming possible. Most individuals or organizations can attune themselves to at least one of these three bodies and work productively with it. This is the form Iowa City has created to enable community programming to occur.

How does this system function? While the cable company was preparing to wire Iowa City, the city council established the position of Broadband Telecommunications Specialist (BTS). As Iowa City's BTS, Mr. Wm. Drew Shaffer is responsible to the City Manager and the BTC for: developing, facilitating, promoting, and coordinating community programming and channels (in cooperation with the cable company's Community Programming Director); monitoring and enforcing the cable franchise and company/city contracts; staffing the BTC; teaching cable-related and production workshops; raising funds; doing research and applications work regarding uses of cable. Headquartered in the public library, the BTS harmonizes the efforts of the cable company, library, city government, and community.

As the cable was being laid, Iowa City constructed a new public library designed to include the city's and library's public access training and production facilities and the cable company's access/local origination studio. A citizen can now learn video production, borrow equipment and tapes, produce and distribute programs over cable, all at the Iowa City Public Library. From the cable company's access center in the library, up to forty hours use of video equipment per user per month is available at no charge, plus sixty minutes of videotape. The library's own access center loans video production equipment to individual users for a maximum of four days per month. A user may also borrow up to two hours of videotape per month free of charge.

Using the Wire: Production

Recognizing that visual literacy is the foundation of effective community programming, the new library maintains facilities and staff which permit a citizen to view movies, videotapes, or slide shows in the library. Or the citizen can borrow these items like books. An Iowa City Library card is the key to contemporary visual literacy and community communication (the cable company also uses a card check-out system at its library-based facility).

Using this key, Iowa Citians have found roles as program producers, directors, writers, actors, camera operators, videotape editors, and lighting technicians. Most people can assume a number of these roles; some produce programs entirely by themselves. Program content, approaches to production, and visual style vary with individual interests, tastes, and experience. No "official" style is required, nor is there a "production code" beyond Federal Communications Commission requirements and a minimal technical level stipulating programs cannot frequently breakup while being cablecast.

Citizens and staffs of each entity produce programming for the library channel, the government channel, the educational channel, and the public access channel. These four channels of community programming were put into operation within one year of cable construction initiation. After three years the government and library channels program twenty-four hours a day; the public access channel has between ten and fifteen hours of original programming produced per week. All involved gain a broader and deeper awareness of their community and the process and products of communication, and some develop their own special mode of expression.

The test of any form or system is how well it functions under changing conditions. A house that looks good in the sunshine may leak in the rain. The city, company, library, and citizens constantly experiment with their cable system to adapt new technology to it and thereby add scope to their communications possibilities.

Taking the Wire One Step Further: Interactive Communications

Influenced by those early interactive experiments in Reading, Pennsylvania, the Iowa City cable system was designed so that the public library, civic government center, and public school system could have two-way audio and video. The city, library, and Access Iowa City (a community organization) in cooperation with the cable company decided to try an experiment in interactive communications that tested new technologies and communications strategies.

Groups of people gathered at four different locations in April, 1982 in Iowa City as part of a regional NFLCP conference to conduct an interactive seminar on the topic of new communications technology. One group gathered at the public library theatre, another at the library studio, a third at the civic center, and one at a community building, also located in the downtown area. They communicated by two-way audio, video, and microcomputer. The members of the four groups determined where the cameras would point, the order of discussion, and how the computers would function. The technical people behind the scenes facilitated the program, they did not direct or limit it.

The experiment required a great deal of technical expertise and willingness by technical talent to accept a new attitude toward production: they had to relinquish control and accept the directions of what used to be called the audience, but who now became the participants. They were able to do this because these technicians shared the public access outlook and often played roles other than that of technicians. Thus, they were more than technicians. The experiment created a new environment in which a new kind of programming could occur; one in which the context and format of television was basically changed, and necessitated a new language and a new way of thinking about television and programming.

The staffs of the company, city, and library worked together technically and administratively. Volunteer technicians were recruited on a large scale and schooled in the new attitude. The engineer for the cable company designed a simple switching system to allow interaction among the locations in the experiment. Local merchants donated television sets so there could be five at each location, allowing everyone to see and hear one another. Microcomputers and microcomputer programmers were found and integrated with the rest of the scenario. ATC, the parent company of the local cable system (Hawkeye Cablevision) provided extra cameras, audio equipment, and technical assistance.

The experiment worked gloriously.

For the first time in Iowa City, citizens controlled television absolutely. "Point the camera at Paula," said a participant. And, lo, the camera was pointed at Paula. The participants expressed their own ideas when they felt it was appropriate, and they were seen and heard at all four locations, and by people watching at home. The pulse of the program rose and fell with the level of participant involvement. No one behind the cameras tried to create an artificial sense of continuity or excitement.

From one location a videodisc machine was plugged into the system and a ballet program was played, illustrating the unique interactive and storage capabilities of that particular piece of technology.

Microcomputers were also used for communication. Graphics tablets were attached to the microcomputers in each location, and the drawings and written comments of the participants also appeared superimposed on the screen over video signals. On one occasion a presenter was flustered when another participant in a different location drew ZZZZZZZ on the graphics tablet which appeared across the screen over his image. One microcomputer programmed another across town and a game was played using the program. The game, a simple one used to initiate the participants in computer use, asked questions such as name, age, and marital status, and an exchange was conducted between those sitting at the keyboards at different locations. Again, everyone saw the exchange. Other participants, using the graphics capability, drew caricatures of each other. For an afternoon Iowa City created and witnessed a rare phenomenon: complete freedom of interactive expression and experimentation on television.

TOWARDS THE FUTURE

Some of the same cooperating entities in this experiment are planning and conducting experiments with interactive computer/cable hookups with the institutional network; designing interactive video/computer programming; conducting a national "Artist and Television" teleconference with New York and Los Angeles via satellite; and communicating internationally via slow-scan with Belfast, Ireland.

The library, schools, city, cable company, and Computer Limited Systems, Inc. (CLSI) have joined in an effort to design an interface so that the library's computer catalog and databanks can be accessed via modem or touch tone phones by the schools on the institutional loop. Eventually home use is also projected.

The city and community producers are designing interactive video/computer programs on topics ranging from public service programs to architecture. One-half inch and three-quarter inch video equipment and microcomputers with built-in light pen or keyboard provide interaction and response capabilities for users.

In October, 1982, ATC, the University of Iowa, community producers, and many others conducted a national satellite teleconference with participants in Iowa City, New York, and Los Angeles. The artist and television was the theme, with guest presenters from around the country. Again, interaction through live performances and discussions with the participants in each location was an integral part of the program.

Community and independent producers, together with company representatives are planning for a slow-scan program with participants in Belfast, Ireland. The topic of the show is music, and the project is an experiment in both technology and television formats.

CONCLUSION

Cable communications in Iowa City, Iowa exemplifies the kaleidoscope of things creating original programming for cable television can be. With the right combination of foresight, planning, designing, and most importantly, people, a truly flexible, usable communications network can be built and used by a great variety of people and organizations; for a variety of purposes. Such a system allows for very simple, practical uses on one level, and at the same time for experimentation with all kinds of new technology and the very constructs of television—out of which unique and very different programming directions, formats, contexts, environments, languages, interactions, needs, and applications may be created. Who knows what one might learn with an Iowa City library card?

RESOURCES

The following persons are knowledgeable about, or have taken part in, the events discussed in this chapter.
1. Bailey, Don, Production Coordinator for the City of Iowa City, 410 E. Washington St., Iowa City, Iowa 52240.

2. Braun, Paul, Manager of Programming Services, American Television and Communications Corp. (ATC), 160 Inverness Drive West, Englewood, Colorado 80112.
3. Buske, Sue, Director of NFLCP, 906 Pennsylvania Ave., S.E., Washington, D.C. 20003.
4. Hindman, Rick, Production Coordinator, Hawkeye CableVision, 546 Southgate Ave., Iowa City, Iowa 52240.
5. Kalergis, Karen, Community Programming Director, Hawkeye CableVision, 546 Southgate Ave., Iowa City, Iowa 52240.
6. Tiffany, Connie, Assistant Director, Iowa City Public Library, 123 S. Linn St., Iowa City, Iowa 52240.
7. Yoder, Todd, Computer Specialist, 530 Ronalds St., Iowa City, Iowa 52240.

Part II.
Creating Original Programming for Cable TV: Avoiding Public Injury and Costly Regulation

*Dee Pridgen
with Eric Engel*

9. Advertising and Marketing on Cable Television: Whither the Public Interest?

Editor's note: Cable television in the 1980s has begun to meet its promise of the 1970s as a medium of abundance—providing far more that mere retransmission of over-the-air broadcast signals, as the chapters in this book testify. A few years ago, most of the new programming developments seemed to be in pay programming. But advertiser supported programming is coming into its own now too. As Dee Pridgen and Eric Engel state in this essay, "Cable television is on the verge of maturing as a conduit for commercial messages and original program services." And they note that "Thus far no blemish has been discovered to mar cable's image as a bonanza of information and entertainment."

Dee Pridgen was Assistant for Special Projects for the Federal Trade Commission's Bureau of Consumer Protection from February 1978 to January 1980. She is currently serving in the Bureau's Division of Advertising Practice.

Eric Engel expects to receive a J.D. from George Washington University's National Law Center in 1982. He received a B.A. with Honors in Communications Studies from the University of California, Los Angeles, in 1979.

This essay originally appeared in CATHOLIC UNIVERSITY LAW REVIEW, *Volume 31, Winter 1982, Number 2.*

Copyright © 1982, The Catholic University of America Press, Inc. Reprinted with permission.

96 / Creating Original Programming for Cable TV

As we explore with the authors the history of the traditional media, however, we become aware of the potential of abuses on cable such as deceptive advertising or unfair marketing practices. The current federal standards which apply to broadcasters do not extend to cable television.

This essay examines the history of the traditional media in these areas and discusses the ways in which the need for federal regulation to protect the consumer can be avoided through responsible self-regulation by the cable industry and independent producers.

The 1980's may be remembered as the decade of cable television. From humble beginnings as a method for improving the broadcast television reception for people in rural or mountainous areas,[1] cable television now brings multiple channels of original programming into over fourteen million homes across the country. With the current penetration level approaching 25% of homes with television, industry analysts are predicting that 50-60% of the homes in the United States will be wired for cable television by the end of the decade.[2] As the coverage of cable spreads, with a corresponding decline in broadcast network television's share of the audience,[3] the advertising community is becoming increasingly interested in exploring the potential of cable television as a distinct marketing medium.

Government agencies traditionally concerned with communications and advertising, such as the Federal Communications Commission (FCC) and the Federal Trade Commission (FTC), are taking a "hands-off" approach to this new medium. Moreover, no federal intervention is likely, absent an

---

1. Indeed, cable television was originally referred to generically as community antenna television, or CATV, because the cable system would set up a large antenna on a hilltop or other location having good reception in order to pick up signals from broadcast stations and redistribute them over cable wires to paying subscribers. *See* FCC INFORMATION BULL. NO. 18, Cable Television 1 (Oct. 1980).

2. *O&M Projects Cable Reaching 60% in '90*, ADVERTISING AGE, Dec. 15, 1980, at 56, col. 4.

3. *J. Walter Thompson's Projections For The Future Of Cable & Commercial TV*, MEDIA INDUSTRY NEWS, Nov. 5, 1980, at 4, predicting a decline in share in prime time network television audience levels from the current 90% to 85% in 1984-85, and 75% in 1989-90.

established record of actual injury to the public interest.[4] Yet must a scandal arise for the creaky wheels of the federal bureaucracy to fashion a response to assure that the "wired nation" will flourish as an honest and open marketplace? By learning from history and planning ahead for coming technological developments, industry, government, and the public can work together to assist the cable medium in becoming an avenue for accurate consumer information without the necessity for heavy-handed regulation.

In Part I of this article, the reality of cable television as an advertising and marketing medium is explored. Advertiser-supported, satellite-distributed cable programming networks already are offering their wares nationwide. Commercials in new formats, such as talk shows or feature films and electronic push-button purchasing from two-way cable television video catalogues, may be introduced soon.

Part II contains an overview of the federal framework for advertising and marketing in the traditional media. In a series of rulings, the FCC has established that broadcasters are responsible for the accuracy of the commercials they disseminate.[5] This background has assisted the FTC in carrying out its mandate to police deceptive advertising.[6] By contrast, the FCC has not imposed similar duties on cablecasters for jurisdictional as well as other reasons.[7] The FTC has also placed some consumer safeguards on direct marketing by conventional means, such as door-to-door and mail-order sales, but has not extended them to newer forms of electronic selling.[8]

In Part III, the industry's efforts to provide a private security force to scrutinize advertising are examined, with a particular emphasis on self-regulation in the broadcasting industry. These private efforts appear to have had a positive effect on the integrity of broadcast television commercials. The initial efforts of the cable industry in this direction appear promising but need more development.

Finally, in Part IV, the possibility of raising, during the local franchising

---

4. While there is no explicit statutory requirement that consumer injury be shown before the FTC will act, as a matter of policy, the agency generally awaits evidence of harm before intervening in the marketplace. *See* FTC, Trade Regulation Rule for the Prevention of Unfair or Deceptive Acts or Practices in the Sale of Cigarettes, 29 Fed. Reg. 8324, 8354-55 (1964), where the Commission specified that "substantive injury to consumers" is one of three factors determining whether a practice could be considered "unfair" under § 5 of the Federal Trade Commission Act, 15 U.S.C. § 45(a)(1) (1958).
5. *See infra* notes 46-73 and accompanying text.
6. *See infra* note 45.
7. *See infra* notes 74-83 and accompanying text.
8. *See infra* notes 84-110 and accompanying text.

(or licensing) process, the issue of the cablecaster's responsibility for the commercials it disseminates is discussed. Unlike broadcast television stations, which are licensed by the FCC,[9] cable television systems receive franchises from local governments. The effect of the first amendment's free press guarantee on a franchise provision concerning advertising and marketing is analyzed. While not diminishing the importance of this consideration, the authors conclude that it would be constitutional for a cable franchisee to be required to assume responsibility for advertising, based both on the cable operator's position as a locally-licensed business and the constitutionally unprotected status of deceptive commercial speech.

In the final analysis, however, it is likely that pressures from local franchising authorities will lead to the development of better self-policing mechanisms for cable television, to the benefit of both the public and the cable industry. Given the sheer number and diversity of local franchising agreements and the rapid pace with which the available franchises are being awarded, it is doubtful that local regulation by itself will be effective. By raising these consumer protection issues during the franchise negotiation process, local citizens can work with the cable industry to create a responsible, yet flexible, mechanism to protect cable viewers from deceptive advertising and abusive marketing schemes.

I. ADVERTISING AND MARKETING ON CABLE TELEVISION: CURRENT STATE OF THE ART AND FUTURE PROJECTIONS

As is rapidly becoming common knowledge, the principal advantage of cable over broadcast television is channel capacity. While only seven VHF broadcast channels can coexist in any one community, a single coaxial cable (about a finger's thickness) currently can deliver up to fifty-two discrete channels.[10] Furthermore, any number of cables can be installed at once, and major "builds" (cable installations) now typically employ at least two cables. Another advantage is cable's unique interactive capacity which allows the viewer to "talk back" to his or her television set.

Considering the competitive positions of cable and broadcast television, it is ironic that cable began as an ally of the broadcast industry.[11] Cable

---

9. 47 U.S.C. § 307 (1976).
10. J. STEARNS, A SHORT COURSE IN CABLE 8 (6th ed. 1981). The number of channels that can be squeezed onto a single cable is expanding almost as rapidly as cable itself. While engineers widely debated the feasibility of 400 mhz technology (capable of delivering 50 to 54 channels) in 1980, 400 mhz technology is now considered commonplace and Magnavox has introduced 440 mhz equipment which it claims can deliver 64 channels on a single cable. *NCTA '81: Hottest Ticket in Mediaville*, BROADCASTING, June 8, 1981, at 44.
11. J. STEARNS, *supra* note 10, at 6.

Ad. & Mktg.—The Public Interest / 99

television originated in the late 1940's and early 1950's in the hills of rural America as a device to improve reception of over-the-air signals from broadcast television stations.[12] A fee was charged for this retransmission service, but the programming and commercial advertising was the same as that on "free" television. Cable television emerged as a separate medium in the mid-1970's when satellite transmission spurred the development of premium, commercial-free "pay cable" services, such as Time, Inc.'s "Home Box Office."[13] Viewers proved willing to pay a subscription fee to see recent and uninterrupted movies at home, a distinct advantage of cable over broadcast television.

In the past few years, however, cable networks have begun to offer special interest programming, partially supported by advertisements. Ted Turner's Cable News Network, for instance, delivers a twenty-four-hour news service as part of the basic cable package offered by many cable operators around the country.[14] This service is not available on broadcast television, but like broadcast programs, it is intended to be advertiser-supported. Similarly, cable programmers appealing to special interest viewers, such as sports fans,[15] women,[16] children,[17] Blacks,[18] Hispanics,[19]

---

12. *Id.*
13. Barrington, *Pay TV: Now a Staple on the Cable "Menu"*, in 2 THE CABLE/BROADBAND COMMUNICATIONS BOOK 1980-1981, at 135-36, 143 (M. Hollowell ed. 1980).
14. Huey, *Future of All-News Network Seems Bright, But Crucial Meeting With Admen Looms*, Wall St. J., Feb. 13, 1981, at 19, col. 4. The network is reaching approximately three million cable homes and has attracted at least 50 advertisers. *Can Ted Turner's Cable News Hang In?*, BUS. WK., Nov. 3, 1980, at 91. One of the nation's largest advertisers, Bristol-Myers, has entered into a 10-year, $25 million contract to sponsor the science and health portions of the infant cable news service. Bronson, *As Marketing Tool of Great Potential Advertisers Begin to Look at Cable TV*, Wall St. J., Dec. 26, 1979, at 11, col. 1.
15. The Entertainment & Sports Programming Network (ESPN), which delivers sports programs to six million cable homes, recently announced a $25 million, five-year advertising contract with Anheuser-Busch, the well-known beer manufacturer. *AB Pours $25 Million into ESPN Contract*, Advertising Age, Nov. 3, 1980, at 4, col. 1. ESPN's president estimated that 70 advertisers would purchase approximately $7 million worth of commercial time from this sports-oriented network in 1980. *Id.*
16. USA Network announced that Bristol-Myers, maker of health and beauty aids, will sponsor a two-hour daily women's show called "Alive and Well." USA Network, Press Release, *Bristol-Myers, USA Network Sign Historic Cable Agreement* (Feb. 18, 1981).
17. USA Network's children's show, "Calliope," has been opened up for commercials deemed "compatible with the entertainment." Christopher, *Alter to Fill the First CTVB Presidency*, Advertising Age, Mar. 2, 1981, at 58, col. 1.
18. Rosenthal, *Cable Advertising Growth Comes In Varied Forms*, TELEVISION/RADIO AGE, May 19, 1980, at 39.
19. *Hispanic TV is Beaming in on the Big Time*, BUS. WK., Mar. 23, 1981, at 122. The SIN Spanish Television Network, distributed by satellite to 79 cable systems, expected to earn $30 million in advertising in 1981. *Id.*

theater buffs,[20] and rock music fans,[21] have announced plans to accept commercials. In fact, of the thirty satellite-distributed cable programming networks announced to date, fully one-half accept advertising.[22] In addition to the satellite services, many cable systems sell advertising on their own channels, using regional "interconnects" to reach larger audiences.[23] The National Cable Television Association, the industry's trade association, is optimistic about the future of commercial support for cable, predicting a tenfold increase in cable advertising revenues, to $350 million, by 1985.[24]

A few cable systems can also boast of "shop-at-home," or direct marketing video services. Modern Satellite Network's "Home Shopping Show" reaches 3.8 million households in forty-seven states.[25] Advertisers can buy nine-minute segments in the half-hour show to demonstrate their wares and to tell viewers how to order.[26] Times-Mirror Satellite Programming Company has joined with Comp-U-Card of America, Inc. to offer a video information and discount buying service to cable viewers,[27] in which the cable system would receive a percentage of the sales originating from its subscribers.[28] The giant of mail-order catalogues, Sears, Roebuck, and Company, is experimenting with the concept of a Video Catalogue Channel to be distributed by cable television.[29]

While cable television seems to be developing into an advertising-sup-

---

20. Rockefeller Center, CBS and ABC all plan to offer culture-oriented pay cable programming, featuring paid advertisements. *RCTV Joins Cable Derby*, Advertising Age, Dec. 15, 1980, at 1, col. 1.

21. Warner-Amex has introduced MTV, a 24-hour rock music network with stereo sound, to attract the elusive, but lucrative, under-35 audience. MTV has prompted inquiries from every major advertising agency in the country. Coeyman, *Cable TV Turns New Ground in World of Commercials*, Christian Sci. Monitor, Apr. 16, 1981, at 11.

22. *Programers Directory*, BROADCASTING, Dec. 15, 1980, at 62-66.

23. A National Cable Television Association survey found that 750 of the 4,700 cable systems in the United States accept advertising for their own channels. *The Cable Rep Entrepreneurs*, MARKETING & MEDIA DECISIONS, Feb. 1981, at 72.

24. *Execs Debate Cable Ads*, Advertising Age, Nov. 24, 1980, at 68, col. 3 (quoting Tom Wheeler, President of the National Cable Television Association). Gerald Hogan, general sales manager for Turner Broadcasting, which offers both "superstation" WTBS Atlanta and the Cable News Network to cable viewers, predicts $500 million in advertising revenues by 1985 for Turner cable operations alone. *Lofty Ad Growth Claims Made by Turner*, Advertising Age, Jan. 19, 1981, at 68, col. 1.

25. Curley, *Formula for a Hit Cable TV Show: Don't Interrupt the Commercials*, Wall St. J., Dec. 30, 1980, at 17, col. 1.

26. Higgins, *Products Star on Cable TV's 'Home Shopping Show'*, Marketing News, Jan. 23, 1981, at 1, col. 1.

27. *The Computer as Retailer*, N.Y. Times, Jan. 9, 1981, at D1, col. 3.

28. *Shop at Home Via Cable and Satellite*, BROADCASTING, Dec. 15, 1980, at 31.

29. *Sears' Wish Book Enters New Video Era*, Advertising Age, May 4, 1981, at 10.

ported medium like its broadcasting counterpart, there appear to be important differences in form and substance looming on the horizon.[30] First, cable's multiple channel capacity has provided a vehicle for special interest programming, offering advertisers a unique opportunity to "narrowcast," or carefully target recipients of their video messages.[31] This fragmentation of the viewing audience may result in an increase in the total number of television advertisements disseminated because different segments will be watching different commercials, rather than a mass audience watching the same advertisement. While the individual viewer may not be seeing a greater number of commercials, the total number of commercials may increase. This trend could have a substantial impact on the workload of a centralized law enforcement agency like the FTC,[32] charged with monitoring deceptive advertising.

Second, cable television's advertising rates are, and probably will continue to be, considerably lower than those of broadcast television.[33] Companies that cannot afford to pay for national broadcast commercial minutes may be enthusiastic purchasers of cable television advertising.[34] Thus, new national television advertisers could emerge, but these new, smaller advertisers may not be familiar with the general requirements that their claims be truthful, nondeceptive, and have a reasonable basis in fact.[35]

---

30. At the Cable Television Administration and Marketing Society's First National Conference on Cable Advertising, Robert Alter, first president of the Cable Television Advertising Bureau, stressed the uniqueness of the medium. Alter said that "[c]able is different. It cannot be forced to fit any existing media mold," and that "[t]he possibilities for developing cable as an advertising medium are probably only limited by our imaginations." *CTVB's Alter Puts "Vive la Difference" Proposition before CTAM Ad Conference*, BROADCASTING, Mar. 9, 1981, at 33.

31. Address by John E. O'Toole, President of Foote, Cone & Belding Communications, Inc., American Association of Advertising Agencies Western Region Meeting (Oct. 18, 1980). *See* Lambert, *Exploring the Potential of Cable Advertising*, BROADCASTING, Feb. 9, 1981, at 16, describing the use of ESPN, a cable sports network, to target successfully potential purchasers of sporting goods store franchises; Poe, *Narrowcasting*, ACROSS THE BOARD, June 1981, at 6.

32. Federal Trade Commission Act, as amended, 15 U.S.C. § 45(a)(1) (1976).

33. Kathryn Creech, Senior Vice President of the National Cable Television Association, has pointed out that the relatively expensive rates of other media is the most important reason for the rise of cable advertising. Rosenthal, *supra* note 18, at 40.

34. *See Burgeoning Role for Cable Sponsors*, Advertising Age, Mar. 2, 1981, at 1, col. 4. Cable "superstation" WTBS Atlanta sells prime-time 30-second spots for $1,100 to $1,800 each, while CBS, ABC, and NBC ask $50,000 to $150,000 for similar time slots. *Cable TV Pitching Big-Spending Advertisers*, Advertising Age, May 5, 1980, at 72, col. 1.

35. A series of FTC cases has held that it is unfair and deceptive for advertisers to make a claim without having substantiation or a reasonable basis in fact before they disseminate the claim. National Dynamics Corp., 82 F.T.C. 488, 543-44 (1973), *modified*, 85 F.T.C. 391

Third, unlike broadcast television, which is governed by an industry code restricting the number of commercial minutes per hour of programming,[36] cable television can offer advertisers more flexible time frames for delivering their messages. In contrast to the thirty-second standard that has evolved in broadcast television, at least one of the new cultural cable programming networks has announced its intention to make time available for two-minute, institutional messages.[37] The possibilities of program-length commercials (referred to as "informercials") and advertiser-produced programming for cable have also been raised. An example of the former occurred on Warner-Amex's QUBE cable system in Columbus, Ohio, when a representative of a local bookstore appeared on "Columbus Alive," a public affairs talk show, to discuss the relative merits of hardcover and paperback books. Four books mentioned during the segment were offered for sale to viewers by the sponsoring bookstore, but at no time was the eight-minute interview identified for what it was—a paid commercial.[38]

Cable systems seem to be suffering from a shortage of programming at present, and it may be that the advertisers themselves will fill that gap with sponsor-produced programs.[39] In this respect, cable resembles the early stages of program development in broadcast television, when advertisers themselves produced the shows that were to be the vehicles for their commercial messages.[40] While there is nothing inherently deceptive about ad-

---

(1975); *In re* Pfizer Inc., 81 F.T.C. 23, 61-62 (1972); Firestone Tire & Rubber, Inc., 81 F.T.C. 398, 427-28 (1972), *aff'd*, 481 F.2d 246 (6th Cir.), *cert. denied*, 414 U.S. 1112 (1973).

36. The National Association of Broadcasters' Television Code limits the amount of nonprogram material to 9 minutes and 30 seconds per hour during prime time and children's programming and to 16 minutes per hour at all other times. NATIONAL ASS'N OF BROADCASTERS, THE TELEVISION CODE 19-20 (21st ed. Jan. 1980) [hereinafter cited as NAB TELEVISION CODE]. Nonprogram material includes billboards, commercials, and promotional announcements. *Id.* § XIV, para. 1, at 18 (Advertising Standards).

37. Herb Granath, head of ABC Video Enterprises, has stated that two-minute commercials may be introduced on ABC's Alpha culture network. *RCTV Joins Cable Derby*, Advertising Age, Dec. 15, 1980, at 1, col. 1. Irving Kahn, Chairman and President of Broadband Communications, has noted that the future of the cable industry "lies not in the 30-second spot but in our extraordinary capacity to develop new means of selling." *Kahn Flies a Red Flag, Donnelly a Green One*, BROADCASTING, Mar. 9, 1981, at 35.

38. The segment cost the Readmor Bookstores about $75. Wicklein, *Wired City, U.S.A.*, THE ATLANTIC, Feb. 1979, at 35, 37. In New York City, a travel agency regularly leases a half-hour slot from Manhattan Cable to air "The First Cable Club Travel Segment," a travelogue promoting the agency's package tours. FTC STAFF, MEDIA POLICY SESSION: TECHNOLOGY AND LEGAL CHANGE (Dec. 31, 1979), at 31 [hereinafter cited as FTC MEDIA POLICY BOOK].

39. For example, Bristol-Myers sponsored and produced the program "Alive and Well." *See supra* note 16.

40. *See* E. BARNOUW, THE SPONSOR 46-58 (1978). Herbert Granath, Vice President of

vertiser-produced programs, or even program-length advertisements, both may raise a consumer protection pitfall because viewers may not be sufficiently aware of the sponsored nature of the programming.

One of the more intriguing marketing opportunities that will distinguish cable from broadcast television is the use of interactive, or "two-way," cable systems to provide instant "push-button" ordering of items offered for sale by means of video catalogues. The most famous of the two-way systems is QUBE, and most of the recent cable franchises awarded around the country include a promise to offer a similar capability.[41] Using the system's interactive capacity, QUBE viewers of a bookstore informercial were invited at the end of the segment to purchase one of the four books offered by pressing an appropriate button on their home consoles.[42] This raises the issue whether consumers could be pressured by this new medium into purchasing products that they would not have bought had they had sufficient time for reflection. On the other hand, the marriage of computerized data bases with interactive cable could turn the television set into a product information library.[43] Comparative price, quality, and warranty information could enrich the purchase decision at minimal cost to the consumer.[44] But who would compile these listings and bear responsibility for their accuracy remains to be seen.

Cable as a distinct advertising medium is beginning to come into its own. The cable networks carry some commercials today and more are expected in the future. Yet cable advertising may differ from conventional television advertising by offering longer formats, narrower audiences, the presence of many new television advertisers, and two-way communication. Given these distinctions, how will the traditional regulatory framework, designed to protect consumers from deceptive advertising and abusive marketing practices, apply? This subject is explored in the next section.

ABC Video Enterprises, has said, "We would like to see advertising agencies and their clients come into this together and possibly produce programs they would sponsor, much like in the early days of television." Poe, *supra* note 31, at 10.
41. *Cable TV: The Race to Plug In*, BUS. WK., Dec. 8, 1980, at 61, 66.
42. Wicklein, *supra* note 38, at 37.
43. *See generally Window on the World: the Home Information Revolution*, BUS. WK., June 29, 1981, at 74.
44. *Cf.* Mitchell, *New Communications Technology: The Prospect for Marketers*, ADVERTISING, Autumn 1980, at 4, describing the United Kingdom's Prestel system which uses specially adapted television sets and ordinary telephone lines to connect viewers with a centralized data base.

II. THE FEDERAL FRAMEWORK FOR ADVERTISING AND MARKETING REGULATION—WHERE DOES CABLE FIT?

The Federal Trade Commission is the major federal agency monitoring advertisements for deceptive practices.[45] Any advertiser disseminating a deceptive commercial over broadcast or cable television is subject to FTC action. However, fulfillment of the FTC's mandate with regard to advertising in the broadcast media has been aided substantially by a series of FCC rulings that hold broadcasters responsible for the accuracy of the commercials they air. Although it is apparent that cable television will also be an advertising medium, the FCC has not applied similar consumer protection measures to cable.

A. *FCC Rulings Establishing Broadcaster Responsibility for Advertising*

The Communications Act of 1934 gives the FCC authority to license broadcasters to serve the "public convenience, interest, or necessity."[46] Television licenses may be granted for a maximum of five years, with renewal at the discretion of the FCC.[47] Under this statutory scheme, the FCC has issued a series of regulations, decisions, and rulings interpreting broadcasters' duty to act in the public interest.[48]

As a general policy, the FCC has avoided content review of allegedly deceptive advertising by referring complaints to the FTC. Indeed, the two agencies have entered into an agreement which provides that the FTC will exercise "primary jurisdiction over all matters regulating unfair or decep-

---

45. Since its inception in the Federal Trade Commission Act of 1914, the Commission has pursued deceptive advertising cases. Pub. L. No. 203, 38 Stat. 717 (1914) (codified at 15 U.S.C. § 45 (1976)). *See, e.g.*, FTC v. Lasso Pictures Co., 1 F.T.C. 374, 378-79 (1919) (advertisements for old movies reissued under new titles must reveal the secondhand nature of the goods). Until the Act was amended in 1938, however, the FTC was limited to a theory that false advertising was an "unfair method of competition." Federal Trade Commission Act, 15 U.S.C. § 45 (1934). The FTC's jurisdiction was expanded in 1938 to include unfair and deceptive trade practices. Wheeler-Lea Act of 1938, 52 Stat. 111 (1938) (currently codified at 15 U.S.C. § 45 (1976)). For an excellent, if now somewhat dated, overview of the law of deceptive advertising, see *Developments in the Law—Deceptive Advertising*, 80 HARV. L. REV. 1005 (1967). *See also* Pitofsky, *Beyond Nader: Consumer Protection and the Regulation of Advertising*, 90 HARV. L. REV. 661 (1977); Pridgen & Preston, *Enhancing the Flow of Information in the Marketplace: From Caveat Emptor to Virginia Pharmacy and Beyond at the Federal Trade Commission*, 14 GA. L. REV. 635 (1980). Many state laws also allow consumers to sue individually if they have been injured by a false or deceptive advertisement. *See* Leaffer & Lipson, *Consumer Actions Against Unfair or Deceptive Acts or Practices: The Private Uses of Federal Trade Commission Jurisprudence*, 48 GEO. WASH. L. REV. 521 (1980).
46. 47 U.S.C. § 307 (1976).
47. *Id.* as amended by Omnibus Budget Reconciliation Act of 1981, Pub. L. No. 97-35, 95 Stat. 736 (1981).
48. *See generally* 47 C.F.R. § 73.99 (1980).

tive advertising in all media, including the broadcast media."[49] The agreement goes on to stress, however, that the FCC "will continue to take into account pertinent considerations in this area [false and misleading advertising] in determining whether broadcast applications for license or renewal of license shall be granted or denied . . . ."[50]

This division of authority does not negate the foundation of FCC directives to broadcasters to the effect that they must:

> assume responsibility for all material which is broadcast through their facilities. This includes all programs and advertising material which they present to the public. With respect to advertising material the licensee has the additional responsibility to take all reasonable measures to eliminate any false, misleading, or deceptive matter . . . . This duty is personal to the licensee and may not be delegated.[51]

The FCC has also charged the broadcast licensee with a continuing obligation "to take reasonable steps to satisfy himself as to the reliability and reputation of prospective advertisers."[52] While "every station must have a program to protect the public in this area,"[53] the FCC has conceded that stations may turn to the National Association of Broadcasters' Code Authority for advice and may also rely on the national broadcast networks' own clearance procedures.[54] The extent of the individual station's clearance program for advertising also "depend[s] upon the size and resources of the station."[55] These FCC rulings demonstrate the Commission's recognition that local stations must be allowed to rely on the self-regulatory

---

49. FCC Public Notice 41503, [Current Service] RAD REG. (P & F) ¶ 11:402 (Apr. 27, 1972) (liaison agreement between FCC and FTC concerning false and misleading radio and television advertising).

50. *Id.*

51. Report and Statement of Policy, re: Commission En Banc Programming Inquiry, July 29, 1960, 20 RAD. REG. (P & F) 1901, 1912-13 (1960) (discussing Report and Statement of Policy at 20 Fed. Reg. 7291 (1960)). *See also* Lee, *The Federal Communications Commission's Impact on Product Advertising*, 46 BROOKLYN L. REV. 463, 484-85 (1980).

52. Alan F. Neckritz, 29 F.C.C.2d 807, 813 (1971).

53. Consumer's Ass'n of D.C., 32 F.C.C.2d 400, 407 (1971). The significance of the network clearance procedure will be discussed in greater detail at *infra* notes 153-57 and accompanying text.

54. *Id.*

55. *Id.* In 1971, the FCC rejected a petition to adopt as part of its rules a "Code of Standards" for television advertising, reasoning that self-regulatory efforts by the broadcasting and advertising industries, as well as the apparent renewed vigor of the FTC, precluded the need for such a code. Adoption of Standards Designed to Eliminate Deceptive Advertising from Television (Petition of TUBE (Termination of Unfair Broadcasting Excesses)), 32 F.C.C.2d 360, 373 (1971).

mechanisms of their network or their trade association to fulfill their responsibility to eliminate deceptive commercials.

FCC licensees are required to take into account FTC rulings on particular commercial messages. If the FTC has issued a final judgment that a particular advertisement is deceptive, the FCC has held that its continued broadcast "would raise serious questions as to whether such stations are operating in the public interest."[56] The FCC has cautioned, however, that:

> licensees should not rely solely on the action or inaction of the Federal Trade Commission, nor should they suspend their own continuing efforts in determining the suitability of advertising material to be broadcast over their facilities. Thus, advertising similar to that found to have been deceptive should raise questions on the part of broadcast stations as to the propriety of such material.[57]

To assist broadcasters in screening for deceptive commercial material, the FTC, in cooperation with the FCC, began sending broadcasters a publication called "Advertising Alert."[58] This bulletin provided notice of the advertisements subject to corrective action by the FTC and also discussed particular problem areas to "familiarize licensees with various deceptive practices so that they will be able to recognize them and take appropriate steps to protect the public against them."[59] Thus, while the FCC has never revoked a broadcast license because the station disseminated misleading advertisements, a series of FCC directives has established that both advertisers and broadcasters have a legal duty to prevent the dissemination of false or misleading commercial messages.[60]

---

56. FCC, Public Notice 41503, [Current Service] RAD. REG. (P & F) ¶ 11:401 (Feb. 21, 1957). The FCC later elaborated this point when it stated:
 Should it come to [the FCC's] attention that a licensee has broadcast advertising which is known to have been the subject of a final Order by the FTC, serious question would be raised as to the adequacy of the measures instituted and carried out by the licensee in the fulfillment of his responsibility, and as to his operation in the public interest.
FCC, Public Notice 11836, [Current Service] RAD. REG. (P & F) ¶ 11:402 (Nov. 7, 1961).
 57. FCC, Public Notice 41503, [Current Service] RAD. REG. (P & F) ¶ 11:401 (Feb. 21, 1957).
 58. FCC, Public Notice 11836, [Current Service] RAD. REG. (P & F) ¶ 11:402 (Nov. 7, 1961) (announcing first issue of the FTC's "Advertising Alert").
 59. *Id.*
 60. There are no legal impediments to carrying out this obligation. The courts have clearly ruled that a broadcaster may refuse to sell time for advertisements it finds objectionable for any reason (with limited exceptions necessary to effectuate the fairness doctrine and to prevent antitrust violations). *See* McIntire v. William Penn Broadcasting Co., 151 F.2d 597, 601 (3d Cir. 1945). *See also* Columbia Broadcasting Sys. v. Democratic Nat'l Comm., 412 U.S. 94, 105 (1973).

The Communications Act itself specifically requires that the sponsors of all paid broadcast material be clearly disclosed.[61] An FCC regulation provides that a single announcement of the sponsor's corporate or trade name, or the name of the sponsor's product, is sufficient provided the product name clearly indicates sponsorship of the message.[62] However, the issue of repeated sponsor identification for longer advertisements has not come up in the broadcast context because the FCC effectively prohibits program-length commercials.[63] "Teaser" broadcasts (short announcements not identifying the advertiser or product, which are to be revealed in subsequent advertisements) are also prohibited under the sponsor identification rule.[64] "Subliminal" advertisements, in which a message is transmitted at levels below the viewing audience's threshold of sensation or awareness, would by definition constitute an evasion of the sponsor identification requirement. The FCC has specifically prohibited this technique for all broadcast material, whether commercial or not.[65]

The FCC has also required broadcasters to take special precautions to assure the adequate separation of program content and commercial messages on children's programs.[66] For instance, a program host or other personality appearing in the program may not promote products on the same children's show on which he or she appears.[67] The FCC has emphasized that, in addition to strictly enforcing sponsor identification of commercial material broadcast during children's programs, it also expects licensees to exercise a higher than average level of care in preventing the broadcast of false, misleading, or deceptive commercials on children's

---

61. 47 U.S.C. § 317(a)(1) (1976).

62. 47 C.F.R. § 73.1212(f) (1980).

63. Public Notice Concerning the Applicability of Commission Policies on Program-Length Commercials, 44 F.C.C.2d 985 (1974). The FCC is apparently more concerned with logging commercials as programs than with the potential for deception and has stated that "[t]he primary test is whether the purportedly non-commercial segment is so interwoven with, and in essence auxiliary to, the sponsor's advertising (if in fact there is any formal advertising) to the point that the entire program constitutes a single commercial promotion for the sponsor's products or services." *Id.* at 986 (footnote omitted). *See also* KCOP-TV, Inc., 24 F.C.C.2d 149 (1970).

64. 40 Fed. Reg. 41935, 41939 (1975) (illustration (F)(32)) (codified at 47 C.F.R. § 73.1212 (1980) without the illustration).

65. Public Notice Concerning the Broadcast of Information by Means of "Subliminal Perception" Techniques, 44 F.C.C.2d 1016 (1974).

66. Petition of Action for Children's Television (ACT) for Rulemaking Looking Toward the Elimination of Sponsorship and Commercial Content in Children's Programming and the Establishment of a Weekly 14-Hour Quota of Children's Television Programs, 50 F.C.C.2d 1 (1974).

67. *Id.* at 16.

shows.[68]

The only other major FCC initiative regarding broadcast advertising concerned the question whether commercial announcements could be subject to the fairness doctrine,[69] which would require broadcasters to provide a reasonable amount of air time to persons holding views opposed to those expressed implicitly or explicitly in the advertisement. In 1967, the FCC found that cigarette advertisements were inherently statements of a point of view in the public controversy over the safety of smoking, and, therefore, broadcasters of cigarette commercials had to make time available to opponents of smoking.[70] The FCC ruling led to an inundation of demands by environmentalists and others for broadcast time to respond to automobile, public utility, and similar ads.[71] The ill-fated experiment ended when the FCC announced that its "cigarette rule" was erroneously based.[72] Currently, the rule is that commercial advertising is not a statement on a controversial issue triggering fairness doctrine obligations as long as the advertisement is restricted to extolling the virtues of the product and takes no explicit position on matters of public controversy.[73]

B. FCC Regulation of Cable Television

In contrast to the precedent of broadcaster responsibility for advertising, the FCC has not placed any obligations on cable systems for the commercials they disseminate, with the single exception of the sponsor identification requirement.[74] FCC regulation of cable television has waxed and waned because the FCC's jurisdiction is indirect. The Communications Act gives the FCC authority over broadcasters and common carriers but

---

68. *See id.* at 18.
69. Under the FCC's fairness doctrine, licensees have a responsibility to devote a reasonable amount of programming time to controversial issues of public importance and to offer a reasonable opportunity for the presentation of contrasting viewpoints. The doctrine was incorporated into the Communications Act by the 1959 amendments, Act of Sept. 14, 1959, Pub. L. No. 86-274, 73 Stat. 557 (codified at 47 U.S.C. § 315(a) (1976)), and upheld by the Supreme Court in Red Lion Broadcasting Co. v. FCC, 395 U.S. 367 (1969).
70. WCBS-TV, 8 F.C.C.2d 381 (1967).
71. *E.g.*, Friends of the Earth v. FCC, 449 F.2d 1164 (D.C. Cir. 1971), holding that the FCC, having applied the fairness doctrine to cigarette commercials, could not refuse to apply the doctrine to automobile and gasoline commercials. *See generally* Note, *Fairness Doctrine: Television as a Marketplace of Ideas*, 45 N.Y.U. L. REV. 1222, 1243-49 (1970); Comment, *Problems in the Application of the Fairness Doctrine to Commercial Advertisements*, 23 VILL. L. REV. 340 (1978).
72. The Handling of Public Issues Under the Fairness Doctrine and the Public Interest Standards of the Communications Act, 48 F.C.C.2d 1, 26 (1974).
73. *See id.* at 25-26.
74. 47 C.F.R. § 76.221 (1980).

does not mention the cable medium.[75] FCC regulation of cable television has been approved only to the extent that such regulation is reasonably ancillary to the effective enforcement of broadcast television regulation.[76] In addition, the FCC does not license cable systems, as it does broadcast stations. Generally, states or municipalities grant cable franchises, and the franchise applicant need only file a pro forma technical document ("signal registration") with the FCC.[77]

Under the "reasonably ancillary" rubric, the FCC issued a series of regulations that, among other things, required larger cable systems to maintain a minimum potential capacity of twenty channels[78] and to designate one or more channels for access by the public and leasing programmers, on a first-come, nondiscriminatory basis, and by local educational and governmental authorities.[79] The Supreme Court refused to sustain these regulations, however, because "[t]he access rules plainly impose common-carrier obligations on cable operators," in violation of the Communications Act.[80] The FCC had earlier dropped its requirement that cable sys-

---

75. The FCC determined early on that cable television operators were not broadcasters under subchapter III of the Communications Act because they did not use the airwaves to transmit signals. Frontier Broadcasting Co. v. Collier, 24 F.C.C. 251 (1958).

76. In the 1960's, the FCC sought to restrict the operation of cable television systems on the theory that their importation of distant signals into a local broadcast television market would harm the local broadcast licensee and would not be in the public interest. Carter Mountain Transmission Corp., 32 F.C.C. 459 (1962); Amendment of Subpart L, Part 91, To Adopt Rules and Regulations To Govern the Grant of Authorizations in the Business Radio Service for Microwave Stations to Relay Television Signals to Community Antenna Systems, 2 F.C.C.2d 725 (1966). The Supreme Court later affirmed the FCC's limited authority over cable, stating that "the authority which we recognize today under § 152(a) is restricted to that reasonably ancillary to the effective performance of the Commission's various responsibilities for the regulation of television broadcasting." United States v. Southwestern Cable Co., 392 U.S. 157, 178 (1968). Under the same theory, the Court later upheld FCC regulations requiring cable systems with 3,500 or more subscribers to originate programming (as opposed to simply retransmitting broadcast signals) and to make facilities available for local production. United States v. Midwest Video Corp. (Midwest Video I), 406 U.S. 649 (1972) (upholding Amendment of Part 74, Subpart K, of the Commission's Rules and Regulations Relative to Community Antenna Television Systems, codified at 47 C.F.R. § 74.1111(a) (1971)). The regulation was first proposed in Inquiry Into the Development of Communications Technology and Services to Formulate Regulatory Policy and Rulemaking and/or Legislative Proposals, 20 F.C.C.2d 201 (1969).

77. 47 C.F.R. § 76.12 (1980).

78. 47 C.F.R. § 76.251 (1972) (systems with 3,500 or more subscribers had to build facilities capable of handling at least 20 channels).

79. *Id.* (governmental and educational users were not to be charged for the use of the facilities; subject to an exception for production costs for live studio presentations exceeding five minutes, public users were not to be charged for use of the facilities).

80. FCC v. Midwest Video Corp. (Midwest Video II), 440 U.S. 689, 701 (1979) (footnote omitted). Section 3(h) of the Communications Act prohibits broadcasters from being treated as common carriers, Communications Act of 1934, Pub. L. No. 416, § 3(h), 48 Stat.

tems originate some programming (as opposed to simply retransmitting broadcast signals),[81] but still applies the fairness doctrine and certain other restrictions to any programming that is originated voluntarily by a cable system.[82] The only advertising restriction the FCC currently applies to such "origination cablecasting" (including cable network programming distributed by satellite) is sponsor identification, as mentioned above.[83] The issue of how to comply with this mandate in the context of longer-format "informercials" has not yet been addressed.

The FCC's relationship with cable television has come full circle in the past two decades. Its original approach was to restrict the expansion of cable in order to maintain the viability of broadcast television, a free service to the public. Subsequently, the FCC saw the cable medium as an opportunity to provide access to television for previously excluded groups such as local community groups and educational interests. With the reversal of the access rules, however, the FCC has taken a pronounced deregulatory approach to the cable industry.

C. FTC Regulations and Federal Statutes Affecting Direct Marketing

Cable television, due to its many channels, special interest programming, and interactive (two-way) capability, is a promising medium, not only for conventional types of advertising, but also for direct marketing efforts. Direct marketing, or (from the consumer's perspective) in-home shopping, encompasses any promotional plan through which a marketer

---

1066 (codified at 47 U.S.C. § 153(h) (1976)), and the Court held this prohibition applicable to cable operators as well. 440 U.S. at 705. The "first-come, first-served" aspect of the access program led to the common carrier characterization by the Court. *Id.* at 701-02. For an excellent analysis of *Midwest Video II*, as well as a good overview of the history of cable regulation, see Note, *Administrative Law—Communications Law—FCC Authority Over Cable Television—FCC v. Midwest Video Corp.*, 1979 WIS. L. REV. 962.

81. Although the Supreme Court approved the origination requirements in *Midwest Video I*, 406 U.S. 649 (1972), the FCC later deleted the requirement, leaving the decision whether to provide original programming to the local system operator. Amendment of Part 76, Subpart G, of the Commission's Rules and Regulations Relative To Program Origination by Cable Television Systems; and Inquiry Into the Development of Cablecasting Services To Formulate Regulatory Policy and Rulemaking, 49 F.C.C.2d 1090 (1974).

82. *See* 47 C.F.R. §§ 76.205, .209, .213, .215 (1980), covering such topics as equal time for political candidates and restrictions on obscenity and lotteries. Even these standards may be lifted under new amendments to the Communications Act. *See* Cable Television Bureau, FCC, Cable Television and the Political Broadcasting Laws: The 1980 Election Experience and Proposals for Change, Report to Sen. Goldwater (Jan. 1981).

83. 47 C.F.R. § 76.221 (1980). The sponsor identification requirement has been waived with respect to classified ads sponsored by individuals, provided that the cable system maintains a publicly-available list of the name, address, and phone number of each advertiser. 47 C.F.R. § 76.221(f) (1980).

Ad. & Mktg.—The Public Interest / 111

sells directly to the customer, without the intervention of a retail outlet.[84] It includes door-to-door sales, telephone solicitations, mail-order catalogues, and television advertisements giving an address or telephone number through which interested persons can place orders. With an interactive cable system, consumers could order directly through the television set by pressing in their credit card or debit card numbers,[85] thus bypassing the extra step of calling or mailing in the order. While direct ordering by means of cable television is farther down the road than cable as a conventional advertising medium, it is nonetheless instructive to examine the current regulatory framework for direct marketing in non-cable media, and to analyze its applicability to cable direct marketing.

In 1972, the FTC promulgated a "cooling-off" rule for door-to-door sales.[86] In essence, the rule provides that any consumer who purchases an item for twenty-five dollars or more from a sales representative at a place other than the normal place of business of the seller has the right to cancel the transaction within three business days of the sale.[87] The Commission based its rule on a finding that personal selling had a record of consumer abuses, such as deceptive door-openers and misrepresentations of price and quality, and that it entailed the nuisance of a visit to the home by an uninvited salesperson.[88] While several states[89] and some FTC cases[90] have

---

84. *See generally* B. STONE, SUCCESSFUL DIRECT-MARKETING METHODS (1979).

85. A "debit" card would allow the holder to initiate an "electronic fund transfer," defined as "any transfer of funds, other than a transaction originated by check, draft, or similar paper instrument, which is initiated through an electronic terminal, telephonic instrument, or computer or magnetic tape so as to order, instruct, or authorize a financial institution to debit or credit an account." 15 U.S.C. § 1693a(6) (Supp. IV 1980). Such a transaction would be the rough equivalent of paying cash or writing a check, as opposed to a credit transaction, which authorizes the debtor to defer payment. 15 U.S.C. § 1602(e) (1976). *See also infra* notes 103-06 and accompanying text.

86. 16 C.F.R. § 429.1 (1973).

87. *Id.* Similar cooling-off provisions had been adopted by a significant number of states. *See* FTC, Cooling-Off Period for Door-to-Door Sales, Trade Regulations Rule and Statement of Basis and Purpose, 37 Fed. Reg. 22,933, 22,935 n.6 (1972). Many states still have such laws on the books. *See, e.g.*, CAL. CIV. CODE §§ 1689.5-.13 (Deering Supp. 1979) (three-day cooling-off, $25 minimum purchase, oral and written notice of right to cancel must be given in the same language used in the sales presentation, seller has 10 days to return down payment and 20 days to pick up cancelled goods); FLA. STAT. ANN. §§ 501.021.-035 (West Supp. 1981) (three-day cooling-off, $25 minimum purchase, seller may keep part of down payment as cancellation fee); GA. CODE ANN. §§ 96-902 to -906 (Supp. 1979) (three-day cooling-off, credit sales only, seller may assess a cancellation fee and pick up fee for cancelled goods even if buyer made no down payment); MISS. CODE ANN. §§ 75-66-1 to -11 (Supp. 1980) (three-day cooling-off, credit sales only, cancellation fee, 40 days to pick up goods, excludes sales on buyer's initiative); N.Y. PERS. PROP. LAW §§ 425-431 (McKinney Supp. 1980-81) (same provisions as California Civil Code).

88. *See* FTC, Cooling-Off Period, *supra* note 87, at 22,937-40 for a summary of the FTC rulemaking record on these points.

applied cooling-off periods to telephone sales, for the most part, this concept has been limited to in-person sales. Sales through television, either broadcast or cable, are not covered.

In the field of unsolicited telephone selling, the abuses of misrepresenting the purpose of the call (e.g., a research survey) and misrepresenting the total price of the contract[91] have been surfacing. The nuisance factor of telephone solicitation has been widely discussed,[92] resulting in several legislative proposals to give telephone subscribers the option of being taken off telephone marketing lists.[93] Computerized, automated dialing systems that deliver prerecorded sales messages have been banned or restricted in several states.[94]

The possibilities for computerized sales interactions via two-way cable television are virtually limitless. A subscriber may respond to a "survey" through his electronic mail system, only to be subjected to a personalized sales pitch delivered from a computer bank of messages based on his an-

---

89. ALASKA STAT. § 45.02.350 (1980); ARK. STAT. ANN. §§ 70-914 to -924 (Supp. 1981); FLA. STAT. ANN. §§ 501.021.-035 (West Supp. 1981); IND. CODE §§ 244.5-2-501 to -505 (Supp. 1981); LA. REV. STAT. ANN. §§ 9:3538-:3541 (West 1981); ME. REV. STAT. ANN. tit. 9-A., §§ 3-501 to -057; tit. 32., §§ 4661-4668 (Supp. 1981-1982); MICH. COMP. LAWS ANN. §§ 445.111-.117 (1981-1982); MONT. CODE ANN. §§ 30-14-501 to -508 (1981); N.D. CENT. CODE §§ 51-18-01 to -09 (Supp. 1979); OR. REV. STAT. §§ 83.710-.750 (1979); WIS. STAT. ANN. §§ 423.201-.205 (West 1974). The three-day cooling-off provisions of Ohio's Home Solicitation Sale Act, OHIO REV. CODE ANN. §§ 1345.21-.28 (Page 1979), have been deemed applicable to telephone sales by the Ohio Supreme Court. Brown v. Martinelli, 66 Ohio St. 2d 45, 419 N.E.2d 1081 (1981).

90. *See* Time, Inc., FTC File No. 781 0003 (Dec. 3, 1980); Budget Marketing, Inc., FTC File No. 782 3015 (Oct. 10, 1980) (consent agreement); Hudson Pharmaceuticals Corp., FTC Docket No. 2860 (July 2, 1980) (consent agreement); Neighborhood Periodical Club, Inc., FTC File No. 712 3119 (Jan. 25, 1980) (consent agreement); Hearst Corp., 82 F.T.C. 218 (1973) (consent agreement); Cowles Communications, Inc., 81 F.T.C. 218 (1972) (consent agreement).

91. *See, e.g.*, Neighborhood Periodical Club, Inc., FTC File No. 712 3119 (Jan. 25, 1980) (consent agreement).

92. *See* Luten, *Give Me a Home Where No Salesmen Phone: Telephone Solicitation and the First Amendment*, 7 HASTINGS CONST. L.Q. 129 (1979); Comment, *Unsolicited Commercial Telephone Calls and the First Amendment: A Constitutional Hangup*, 11 PAC. L.J. 143 (1979).

93. *See* discussion of 1978 California legislative proposal in Comment, *Unsolicited Commercial Telephone Calls*, *supra* note 92, at 159-60.

94. California, Florida, and Maryland ban all use of automatic dialing devices. CAL. PUB. UTIL. CODE § 2872 (West Supp. 1981); FLA. STAT. ANN. § 365.165 (West Supp. 1981); MD. ANN. CODE art. 78, § 55C (1980). Alaska and Wisconsin ban all "junk telephone calls" without the prior written consent of the consumer. ALASKA STAT. § 45.50.472 (1980); WIS. STAT. ANN. § 134.72 (West Supp. 1981-1982). Virginia prohibits the use of devices that do not disconnect when the consumer replaces the receiver. VA. CODE § 18.2-425.1 (Supp. 1981). In Illinois, automatic dialing systems may only be used with a live operator. Ill. Com. Comm'n Docket 0087 (1978).

swers to the survey. Overreaching sales techniques and potential invasions of privacy could provoke a call for extension of consumer protection measures to interactive cable television. In addition to the constitutional issues associated with attempts to restrict nondeceptive commercial speech, it is unclear whether electronic sales would give rise to the same level of consumer injury as door-to-door, in-person sales. Measures such as cooling-off periods, subscriber option to be taken off call lists, and clear identification of the purpose of the communication at its outset may have to be considered in the future but should be approached cautiously so as not to stifle innovation in this new medium, which has the potential to provide a wealth of valuable information to consumers.

Another issue in direct marketing, whether conventional or electronic, is the amount of time the seller should take to send the ordered merchandise. In 1975, the FTC issued a regulation (the Mail-Order Rule) which in essence provided that sellers should ship merchandise ordered by mail to buyers within the promised time period or within thirty days if no time had been specified.[95] If the seller is unable to send the goods within the applicable time period, the buyer must be given the option to consent to a delay for a specified time or to cancel the order and receive a prompt refund.[96] The FTC rulemaking record contains over 20,000 pages of consumer complaints regarding mail order sales, with failure to ship prepaid merchandise by far the most frequent complaint.[97] Despite the enforcement efforts of the FTC, the United States Postal Service[98] and the industry trade association,[99] late delivery and nondelivery remains a significant consumer problem.[100]

Despite an expected upsurge in direct sales by telephone[101] and other electronic media, including cable television, neither the FTC's Mail-Order Rule nor the Postal Service statutes apply to transactions other than those

---

95. 16 C.F.R. § 435.1(a)(1) (1976).
96. *Id.* at § 435.1(b)(1).
97. Mail Order Merchandise, Promulgation of Trade Regulation Rule, 40 Fed. Reg. 51,582 (1975).
98. The Postal Service has jurisdiction over mail-order problems involving criminal fraud, 18 U.S.C. § 1341 (1976), and misrepresentations of fact by the seller, 39 U.S.C. § 3005 (1976).
99. The Direct Mail/Marketing Association is a 3,400-member industry group that helps resolve consumer mail-order problems. It currently receives 20,000 to 30,000 mail-order complaints a year, most involving nondelivery. *See* U.S. GENERAL ACCOUNTING OFFICE, THE FEDERAL TRADE COMMISSION'S MAIL ORDER RULE NEEDS IMPROVED MONITORING AND ENFORCEMENT 3 (Jan. 19, 1981).
100. *Id.* at 2. *See also Federal Study Advising Mail-Order Rule Reform*, N.Y. Times, Jan. 26, 1981, § A, at 24, col. 5.
101. GAO Report, *supra* note 99, at 1.

where the consumer orders by mail.[102] Thus, they would not apply to cable television marketing efforts, unless the buyer uses the mails to order rather than the telephone or the interactive feature of a cable system. Whether late shipment protection should be extended to the electronic media remains an unanswered question.

To the extent that consumers use electronic fund transfer (EFT)[103] rather than credit to purchase items marketed on cable, some significant consumer rights would be lost. The Fair Credit Billing Act allows credit card customers to use the creditor's dispute resolution procedure. In turn, it requires the creditor to withhold charges for goods or services which the consumer did not accept or which were not shipped as agreed.[104] Furthermore, under certain circumstances, the consumer can withhold payment from the credit card issuer by asserting a claim or defense against the merchant regarding the quality of the goods purchased.[105] The Electronic Fund Transfer Act, by contrast, was apparently aimed primarily at the bank-depositor relationship and does not contain these marketing-oriented, consumer protection measures.[106] An EFT purchase would operate in essentially the same way a cash purchase does, since payment could not be withheld if the goods were not sent or were defective.

Finally, it is unclear how consumers will obtain presale warranty information for goods marketed directly on cable television. Congress passed the Magnuson-Moss Warranty Act in 1975 with the stated goal of improving the adequacy of warranty information available to consumers, thereby improving competition in the marketing of consumer goods.[107] Under an

---

102. 16 C.F.R. § 435.1(a)(1) (1981) states that the FTC Mail-Order Rule is applicable to "any order for the sale of merchandise *to be ordered by the buyer through the mails.*" (emphasis added). The Postal Service Statute dealing with false representations (as opposed to mail or wire fraud) requires that there be "a scheme or device for obtaining money or property *through the mail* by means of false representations" in order to constitute a violation. 39 U.S.C. § 3005(a) (1976) (emphasis added).

103. Most development of EFT systems thus far has been in banking rather than in marketing. If "home banking" and "home shopping" services develop simultaneously, it is likely that the cable medium could become a major stimulus for the use of EFT. There may be a regulatory stumbling block to this development, however, because paper receipts currently must be made available at all EFT terminals. Electronic Fund Transfer Act of 1978, 15 U.S.C. §§ 1693a(7), 1693d(a) (Supp. II 1978). *See generally* Broadman, *Electronic Fund Transfer Act: Is the Consumer Protected?*, 13 U. OF S.F.L. REV. 245 (1979).

104. 15 U.S.C. §§ 1666(a), (b)(3) (1976).

105. *Id.* § 1666i (transaction must exceed $50 and occur in the same state or within 100 miles of the cardholder's address).

106. *See generally* 15 U.S.C. §§ 1693-1693r (Supp. II 1978); 12 C.F.R. §§ 205-205.14 (1981). For a summary of the features and failings of the Electronic Fund Transfer Act, see Broadman, *supra* note 103, at 13.

107. Magnuson-Moss Warranty—Federal Trade Commission Improvement Act, Pub. L. No. 93-637, 88 Stat. 2183 (codified at 15 U.S.C. §§ 2301-2312a (1976)).

FTC regulation implementing the Act, a written copy of the warranty (if any is offered) on all consumer products priced over fifteen dollars must be available in retail stores prior to any sale.[108] As for mail-order catalogues, either the full text of the warranty must be reported, or the catalogue must disclose that a free copy of the written warranty can be obtained on request.[109] Due to a rather broad definition of "mail-order,"[110] it appears that telephone and interactive cable sales would be included. The capability of interactive cable television to provide instant access to a central data base would clearly benefit consumers in obtaining the warranty information through the cable medium itself, rather than through the more cumbersome procedure of requesting the warranty by mail.

In marketing, as in advertising, a regulatory structure has been erected to curb abuses in the traditional media; these safeguards, for the most part, are not applicable to cable marketing. Yet federal regulation is unlikely unless a record of consumer injury develops. If abuses in this new use of cable television are to be prevented, the most likely stimulus at this time would be either industry self-regulation or local franchising agreements. Each of these possibilities is discussed in turn in the next two sections.

III. INDUSTRY SELF-REGULATION OF ADVERTISING AND MARKETING—PRECEDENTS FROM BROADCASTING AND THE PROSPECTS FOR CABLE

Advertising and marketing in the broadcast media, particularly television, is filtered through a series of self-regulatory standards and procedures to assure its accuracy. The National Association of Broadcasters (NAB) has developed a fairly detailed Television Code[111] and issues detailed guidelines and Code interpretations as needed.[112] The NAB also prescreens commercials in a few sensitive categories.[113] The three major

---

108. 16 C.F.R. § 702.3 (1981).
109. 16 C.F.R. § 702.3(c)(2)(i) (1980).
110. The regulation defines "catalogue or mail-order sales" to include "any solicitation for an order for a consumer product with a written warranty, which includes instructions for ordering the product which does not require a personal visit to the seller's establishment." *Id.*
111. NAB TELEVISION CODE, *supra* note 36. Sections I through VIII are concerned with program standards. Sections IX through XV deal with advertising standards.
112. *See, e.g.*, NAB CODE AUTHORITY, CHILDREN'S TV ADVERTISING GUIDELINES (2d ed. Apr. 1977).
113. The NAB Code Authority prescreens all broadcast commercials in the following four areas: (1) children's toys, (2) children's premiums, (3) personal care products, and (4) margarine and vegetable oil products involving health claims. LaBarbera, *Analyzing and Advancing the State of the Art of Advertising Self-Regulation*, 9 J. OF ADVERTISING 27, 30 (1980).

broadcast networks preview all commercials aired, subjecting them to an internal audit under standards for accuracy and "taste."[114] The Association of National Advertisers (ANA), the trade association of major national advertisers, and the American Association of Advertising Agencies (4A's) created the National Advertising Division (NAD) of the Council of Better Business Bureaus to review advertisements of questionable veracity.[115] This rather tightknit web appears to have helped prevent, at least on national television, the dissemination of blatantly fraudulent advertising. The cable industry, by contrast, is just beginning to explore the need for self-regulation in this burgeoning new medium. The development of internal regulation in broadcast television and its potential applicability to cable is examined next.

A. The NAB and the Television Code

The NAB had its origins in the radio industry, having had its first organizational meeting in 1923.[116] After the creation of the Federal Radio Commission (precursor to the FCC) in 1927,[117] criticism of certain radio industry practices, such as the playing of unidentified phonograph records and, curiously, direct advertising (today called commercials), began to mount.[118] In an attempt to forestall further government regulation, the NAB Board of Directors approved the organization's first Code of Ethics in 1928.[119] This first effort was rather general,[120] but the following year, the NAB revised the Code, with six out of the seven principles of conduct addressing consumer deception and safety issues.[121]

The Code provisions were more honored in the breach than otherwise by most radio stations, however, leading the Federal Radio Commission to threaten "proper legislation" (including the possibility of nationalizing the industry) if the broadcasters did not eliminate "false, deceptive or exagger-

---

114. J. PRICE, THE BEST THING ON TV 127-28 (1978). For a discussion of CBS' screening practices, see Consumers Ass'n of D.C., 32 F.C.C.2d 400, 401 (1971).
115. LaBarbera, *The Shame of Magazine Advertising*, 10 J. OF ADVERTISING 31, 36 (1981). The NAD reviews both broadcast and print advertisements.
116. D. Mackey, The National Association of Broadcasters—Its First Twenty Years, 8 (Aug. 1956) (unpublished Ph.D. thesis in NAB Library) [hereinafter cited as Mackey].
117. Radio Act of 1927, Pub. L. No. 69-632, 44 Stat. 1162 (repealed 1934).
118. Mackey, *supra* note 116, at 346.
119. *Id.*
120. In his doctoral thesis on the history of the NAB, David Mackey said of the 1928 version of the Code: "It's obvious that in its efforts to keep everyone happy, the ethics committee came up with a Code which not only had no teeth, but very soft gums." *Id.* at 350.
121. The full text of these six principles was:
   SECOND. When the facilities of a broadcaster are used by others than the owner, the broadcaster shall ascertain the financial responsibility and character of such

ated" advertising statements.[122] The FTC, on the other hand, proposed a conference with the industry "for the purpose of cooperatively drafting rules by which the industry could regulate itself and thus avoid the pitfalls of fraudulent or misleading advertising."[123] While the Radio Commission ultimately concluded that government ownership was impractical and that regulation of commercial advertising would impair the quality of programming,[124] the Association of National Advertisers three years later proposed a plan to eliminate undesirable advertising by "voluntary internal censorship" with "the hope of substituting 'self-regulation' of advertising for projected government regulation . . . ."[125]

From the beginning of commercial television, the radio Code of Ethics was considered applicable to television broadcasters, but in 1950 the NAB announced plans to develop a separate television code to deal with television's visual aspects.[126] As was the case in the development of the radio code, one of the likely motivations was to fend off government regulation, a threat apparently believed to be possible but not imminent.[127] When Senator William Benton introduced a Senate resolution calling for the cre-

client, that no dishonest fraudulent or dangerous person, firm or organization may gain access to the Radio audience.
   THIRD. Matter which is barred from the mails as fraudulent, deceptive or obscene shall not be broadcast.
   FOURTH. Every broadcaster shall exercise great caution in accepting any advertising matter regarding products or services which may be injurious to health.
   FIFTH. No broadcaster shall permit the broadcasting of advertising statements or claims which he knows or believes to be false, deceptive or grossly exaggerated.
   SIXTH. Every broadcaster shall strictly follow the provisions of the Radio Act of 1927 regarding the clear identification of sponsored or paid-for material.
   SEVENTH. Care shall be taken to prevent the broadcasting of statements derogatory to other stations, to individuals, or to competing products or services, except where the law specifically provides that the station has no right of censorship.
Excerpts from the 1929 Code, app. 2, *reprinted in* L. WHITE, THE AMERICAN RADIO: A REPORT ON THE BROADCASTING INDUSTRY IN THE UNITED STATES FROM THE COMMISSION ON THE FREEDOM OF THE PRESS (1947).
   122. *Warning Issued on Blatant Advertising; Commission Proposes Self-Regulation to Stave Off Congressional Action, Upholds American System*, BROADCASTING, Jan. 1, 1932, at 12. The Commission expressed concern about offensive programming as well as questionable commercial material. *Id.*
   123. *U.S. Trade Body Head Lauds Radio Ethics*, BROADCASTING, Mar. 15, 1932, at 5.
   124. *Radio Board Cold to Advertising Cut*, N.Y. Times, June 10, 1932, at 12, col. 3.
   125. *ANA Plans to Bar False Advertising; Proposes Self-Regulation to Obviate Federal Control*, BROADCASTING, Jan. 15, 1935, at 26.
   126. *T.V. Standards; NAB to Set Up Code Unit*, BROADCASTING, May 1, 1950, at 50.
   127. An editorial in Telecasting (a subsection of BROADCASTING) noted:
   It was probably inevitable at this stage of television development that there would have occurred enough lapses in common sense and good taste in programming to arouse a fear within the industry that unless formal corrective action were taken at once, censorship was just around the corner.

ation of an eleven-member National Citizens Advisory Board for Radio and Television in 1951,[128] however, a trade press editorial condemned the proposal as an attempt at "regulation by lifted eyebrow."[129] Suddenly, the promulgation of a television code became a high priority for the National Association of Radio and Television Broadcasters.[130] Before the end of 1951, the first television code was adopted. *Broadcasting* magazine noted that the "stringent code" was approved by the industry "with the eyes of Congress upon them."[131] As passed, the Code sought to remind telecasters that "they must be choosy in admitting advertisers to their facilities as well as careful to require truth and consideration in commercial messages . . . ."[132]

The Television Code has undergone twenty revisions since its first adoption, but the basic principles have remained constant. It acknowledges that broadcasters are ultimately responsible for all material broadcast by their stations, including advertising.[133] Broadcasters are advised to "refuse the facilities of their stations to an advertiser where they have good reason to doubt . . . the truth of the advertising representations."[134] As a general rule with regard to advertising, the Code states:

> great care [should] be exercised by the broadcaster to prevent the presentation of false, misleading or deceptive advertising. While it is entirely appropriate to present a product in a favorable light and atmosphere, the presentation must not, by copy or demonstration, involve a material deception as to the characteristics, performance or appearance of the product.[135]

More specific guidance on particular types of advertising is provided. For instance, commercials directed primarily toward children must be clearly separated from the program[136] and must not feature personalities

---

Television Code I—an Editorial, Telecasting, May 1, 1950 at 4 *cited in* BROADCASTING, May 1, 1950, at 50.
128. S.J. Res. 76, 82d Cong., 1st Sess., 97 CONG. REC. 6117 (1951).
129. *Bye Bye, Bill of Rights*, BROADCASTING, Sept. 10, 1951, at 58.
130. In 1951, the NAB merged with the Television Broadcasters Association and changed its name to the National Association of Radio and Television Broadcasters (NARTB). The NARTB said that the Television Code would show the public that the industry "means business." *T.V. Code Takes Shape*, BROADCASTING, Oct. 8, 1951, at 71. The organization reverted to the name NAB in 1958.
131. *Stringent TV Code*, BROADCASTING, Oct. 22, 1951, at 23.
132. *Id.*
133. NAB TELEVISION CODE, *supra* note 36, at 1 (Preamble). The preamble goes on to state that the responsibility is shared with television advertisers. *Id.*
134. *Id.* § IX, para. 1(B), at 11 (Advertising Standards).
135. *Id.* § X, para. 1, at 15.
136. *Id.* § IX, para. 6(B), at 13.

or cartoon characters regularly appearing in the sponsored program.[137] "Bait and switch" advertising is specifically prohibited,[138] as are program-length commercials.[139] Personal endorsements must be genuine and based on personal experience.[140] References to research, surveys, or tests must not "create an impression of fact beyond that established by the work that has been conducted."[141] The use of an actor in a white coat (i.e., simulating a medical professional) to recommend a health product is not permitted.[142] Contests and premium offers should spell out all details and not exaggerate the prizes to be won.[143]

The NAB Television Code, to which 67% of United States television stations and the three national networks subscribe,[144] is administered by the Television Code Authority.[145] This group issues a continual series of "interpretations" of the Code for the guidance of subscribers. For instance, although mail order merchandising is nowhere specifically mentioned in the Television Code, the Code Authority has interpreted the general provisions regarding the integrity and truthfulness of advertisers to advise that "mail order advertising, because it exhorts the viewer to invest his money in a product sight unseen, requires greater vigilance by the broadcaster than is normally required."[146]

The Code is enforced formally by control over the use of the "NAB Television Seal of Good Practice,"[147] and by the Code Authority's bi-yearly monitoring of each television station.[148] The NAB has rarely sus-

---

137. *Id* § IX, para. 6(E), at 13.
138. *Id* § IX, para. 14, at 14. Bait and switch involves "goods or services which the advertiser has no intention of selling . . . offered merely to lure the customer into purchasing higher-priced substitutes." *Id*
139. *Id* § IX, para. 15, at 15.
140. *Id* § IX, para. 6, at 15.
141. *Id* § X, para. 2, at 15.
142. *Id* § XI, para. 4(A), at 17.
143. *Id* §§ XII & XIII, at 17-18.
144. NAB, LEGAL GUIDE TO FCC BROADCAST RULES, REGULATIONS AND POLICIES, ch. IV, at 2 (May 1977).
145. The NAB Code Authority employs a full time staff of 34 and has offices in New York City, the District of Columbia, and Los Angeles. This group preclears about 3,000 advertisements per year. The Code Authority also has a medical and scientific panel which provides free advisory service to NAB staff. Interview with Jerry Lansner and William Schulte, NAB Code Authority Staff, New York, N.Y. (Feb. 23, 1981) [hereinafter cited as Lansner Interview].
146. *Mail Order Advertising Requires Special Care in Evaluating Offer Before It is Aired*, NAB TV CODE NEWS, Dec. 1966, at 3.
147. *See* NAB TELEVISION CODE, *supra* note 36, § III(4), at 25 ((Regulations and Procedures). Section III(4) states that authority to use the Seal may be revoked or suspended by a vote of two-thirds of the Television Board of Directors.
148. NAB CODE AUTH., BROADCAST SELF-REGULATION: WORKING MANUAL (1976 in-

pended a station's authority to use the Seal;[149] however, there may be enough public awareness to make the threat of withdrawal a fairly effective deterrent.[150] Most of the policing appears to be accomplished through advice and comment behind the scenes. Member stations can and frequently do call the Code Authority for advice.[151] Advertising agencies may also come in with questions in the course of preparing a campaign. After a commercial is aired, advertisers may file complaints with the NAB regarding the claims of their competitors. These questions can be negotiated with the member stations who broadcast the commercials in question.[152] Another major influence on compliance is the clearance procedure for commercials employed by the broadcast networks; this procedure will be discussed in the next section.

B. Broadcast Television Network Internal Review Procedure

While little has been formally documented about the rules and procedures for network clearance of commercials, there can be no doubt that it plays a pivotal role in maintaining the honesty of advertising shown on national network television. The networks review *every* commercial before it is aired, a total of approximately 50,000 per year.[153] The networks are prominent members of the NAB, each having a permanent seat on the Board of Directors. The broadcast networks consider applicable NAB Code standards when they preview advertisements.[154] Thus, many com-

---

sert) [hereinafter cited as BROADCAST SELF-REGULATION]; Lansner interview, *supra* note 145. During this review of the station's log books, certain types of commercials are generally flagged for additional review. *Id.*

149. One incident occurred in the early 1960's when 65 subscribers withdrew or were ejected from subscription to the Code because of a dispute over advertising for a hemorrhoid medicine. Nearly all subsequently rejoined the fold. BROADCAST SELF-REGULATION, *supra* note 148, *Commercial Clearance* (1972 insert).

150. The NAB has pledged to publicize the Seal and its significance. NAB TELEVISION CODE, *supra* note 36, § III(2), at 25 (Regulations and Procedures). In 1967, the NAB commissioned a Roper survey of the public's awareness of the Seal and found that 54% of those asked had seen the Seal before, and that most of those who had seen it were aware of the basic provisions of the Code. *Special Report: Research Shows Public Prefers Self-Regulation*, NAB TV CODE NEWS, April 1967, at 1. Only one-third believed the industry had imposed the rules on itself, with most crediting the FCC or some other government agency. *Id.*

151. Lansner interview, *supra* note 145.

152. *Id.*

153. *Id.* In 1971, the CBS Television Network Program Practices Department alone had a staff of 34 in New York and Los Angeles. Advertisements are usually submitted in script or storyboard form. All product claims must be substantiated, and the network staff, the advertiser, and/or the advertising agency commonly discuss the substantiation or presentation of a claim before it is accepted by the network. Consumers Ass'n of D.C., 32 F.C.C.2d 400, 401 (1971).

154. Lansner Interview, *supra* note 145.

mercials containing potential Code violations are never broadcast, but are modified or rejected at the network level. Indeed, other self-regulatory bodies that deal with television advertising have been greatly assisted by the network procedures.[155]

Network clearance, which grew partially out of a desire to preserve the licenses of affiliated stations, was formalized during the early 1970's.[156] During this time, the three networks also began a practice, still continued today, of forwarding the scripts or storyboards of all commercials aired to the FTC's Advertising Monitoring Unit.[157] Thus, the FTC is assisted in its consumer protection mission by the network clearance procedures and by having scripts of commercials made available for review.

C. *The National Advertising Division of the Council of Better Business Bureaus*

The National Advertising Division (NAD) and its appellate body, the National Advertising Review Board (NARB), were sponsored in 1971 by the major advertising associations (the ANA and the 4A's) under the aegis of the Council of Better Business Bureaus.[158] The NAD does not administer a code as such, but its stated goal is "to achieve and sustain high standards of truth and accuracy in national advertising."[159]

The NAD does not preview commercials but resolves complaints from competitors and consumers. It also initiates almost half of its cases by monitoring advertisements on radio and television and in magazines and newspapers.[160] Cases that cannot be resolved by the NAD are reviewed by the NARB, a court of appeals composed of ten five-member panels of rep-

---

155. NAB Code Authority staff confirmed that they would have to expand their operations greatly if the networks stopped clearing commercials. *Id.* Similarly, Lorraine Reed, Senior Vice President of the Council of Better Business Bureaus' National Advertising Division (NAD), has said that while 53% of their cases are from the print media, only 32% are from television. The network prescreening procedures, therefore, probably help keep misleading advertisements from being aired. Interview with Lorraine Reed, New York, N.Y. (Feb. 23, 1981) [hereinafter cited as Reed Interview].
156. Interview with Alfred Schneider, Vice President of Incorporation, American Broadcasting Company, New York, N.Y. (Feb. 23, 1981).
157. Letters from Miles W. Kirkpatrick to American Broadcasting Co., Columbia Broadcasting System, and National Broadcasting Co. (Mar. 22, 1971). Prior to 1971, the FTC required submission of only periodic samples of network advertising scripts. FTC, *How FTC Monitoring System Operates, Advertising Alert No. 2* [Current Service] RAD. REG. (P & F) ¶ 11:402 (Feb 12, 1962).
158. LaBarbera, *The Shame of Magazine Advertising, supra* note 115, at 36.
159. LaBarbera, *Advertising Self-Regulation, supra* note 113, at 30. The NAD, however, does distribute specific guidelines for children's advertising. *NAD Reviews 1980 Accomplishments*, NEWS FROM NAD, Jan. 15, 1981, at 1.
160. Reed Interview, *supra* note 155.

resentatives from advertisers, advertising agencies, and the public.[161] While this system has been in existence for over eight years, only thirty-five panel decisions regarding truth and accuracy in national advertising have been rendered to date.[162] Referral to the FTC is considered the ultimate sanction, but thus far this has not proven necessary.[163]

The work of the NAD/NARB apparently has been very effective. Former FTC Chairman (now Commissioner) Michael Pertschuk has praised the group for having "skimmed the cream of deceptive ads, outrageous frauds and misrepresentations. Thanks to you, the latter-day ancestor of the silver tongue snake oil purveyor has been tongue-tied."[164] Former FTC Commissioner Robert Pitofsky, who was Director of the Bureau of Consumer Protection at the time the NARB was created, has stated:

> [T]he NARB is as successful an effort at self regulation as any we have witnessed in the country. Many of the cases that might otherwise vex federal enforcers are nipped through the self regulation process, and policing now occurs inside advertising's house and therefore without the inevitable friction of government regulation of private activities.[165]

D. Cable Advertising and the Future of Self-Regulation

The cable industry begins its emergence as an advertising vehicle at a time when deregulation has become fashionable. Yet, self-regulation can have a positive, prophylactic effect. Will this relatively young medium act responsibly to ensure that cable television is not handicapped from the start by unsubstantiated or misleading advertising banned by the more established media? The signs thus far are encouraging, although more could be done to bring cable up to the level of broadcast television in terms of advertising self-regulation.

The National Cable Television Association (NCTA), the cable counterpart to the NAB, has formed a committee to study advertising self-regulation.[166] The NAB Code appears to be the primary model, although the

---

161. Zanot, *A Review of Eight Years of NARB Casework: Guidelines and Parameters of Deceptive Advertising*, 9 J. OF ADVERTISING 20 (1980).
162. *Id.*
163. LaBarbera, *Advertising Self-Regulation*, *supra* note 113, at 31.
164. Address by Chairman Michael Pertschuk, FTC, Annual Meeting of the National Advertising Review Board, New York, N.Y. (Nov. 8, 1978).
165. Address by Commissioner Robert Pitofsky, FTC, National Convention of the American Advertising Federation, Washington, D.C. (June 12, 1979).
166. Statement by Kay Koplovitz, President of the USA Network, The Consumer and Cable Television: A National Conference, Washington, D.C. (Feb. 27, 1981). *See also Can Advertising Regulate Itself?*, MARKETING & MEDIA DECISIONS, July 1981, at 37.

NCTA is taking a cautious approach.[167] Some aspects of the NAB Code, such as the limits on commercial minutes per hour, may not be deemed appropriate for cable, which can offer advertisers greater flexibility due to the large number of channels available.[168] In addition, absolute bans on commercials for certain types of products (e.g., liquor and contraceptives)[169] may not be needed in the cable medium, which has the technical capacity to allow parents to prevent their children from having access to "adult" channels. Indeed, any attempt to restrain the free flow of *truthful* commercial speech on cable could be viewed as anticompetitive.[170]

NCTA is not only looking back to the NAB Code, but is also reportedly studying the self-regulatory code of a more futuristic and experimental medium,[171] the computerized information retrieval system called Prestel, now operating in the United Kingdom.[172] Prestel is a "viewdata" or "videotex" system, which in essence links the user's television set to a central data bank via telephone lines (although the same effect could be achieved directly through the coaxial cable used in cable television). Prestel gives the user access to over 160 information providers, including Reuters, British Rail, FINTAL (a commercial publisher of financial information) and American Express.[173] The system, developed by the British Postal System, is operated as a common carrier. The information providers (not Prestel) are solely responsible for the content they disseminate, which includes

---

167. Telephone interview with Char Beales, Vice-President for Media Services and Research, National Council of Television Advertisers (Feb. 12, 1981).
168. *See* NAB TELEVISION CODE, *supra* note 36, §§ XIV-XV, at 18-23 (Advertising Standards). The NAB limitation of the supply of commercial time has been challenged by the United States Department of Justice as an antitrust violation. United States v. National Ass'n of Broadcasters, No. 79-1549 (D.D.C. filed June 14, 1979). For a discussion of the economic effects of such restrictions, see FTC MEDIA POLICY BOOK, *supra* note 38 at 124-40; and H. BEALES, TELEVISION PROGRAM QUALITY AND THE RESTRICTION ON THE NUMBER OF COMMERCIALS (FTC Bureau of Economics Working Paper No. 30, June 1980).
169. *See generally* NAB TELEVISION CODE, *supra* note 36, § IX (Advertising Standards).
170. *Cf.* American Medical Ass'n, 94 F.T.C. 701 (1979), *enforced as modified*, 1980-2 Trade Cas. (CCH) ¶ 63,569 (2d Cir. Oct. 7, 1980), in which the FTC struck down the AMA's near-total prohibition of advertising by physicians. Because self-regulation is permissible under the rule of reason if it promotes competition, the FTC's decision in the AMA case allowed that group to issue and enforce "reasonable ethical guidelines" to weed out false or deceptive advertising by member physicians. 94 F.T.C. at 1037. *See also* Address by E. Perry Johnson, Director of the FTC Bureau of Competition, 17th Annual Symposium on Trade Association Law and Practice of the Antitrust Law Committee, Bar Association of the District of Columbia, Washington, D.C. (Feb. 25, 1981).
171. Beales Interview, *supra* note 167.
172. For a discussion of Prestel and similar videotex systems and their capabilities, see *Window on the World: The Home Information Revolution*, BUS. WK., June 29, 1981, at 74. *See also* R. WOOLFE, VIDEOTEX: THE NEW TELEVISION/TELEPHONE INFORMATION SERVICES (1980).
173. FTC MEDIA POLICY BOOK, *supra* note 38, at 52.

some advertisements.[174]

While advertising is not regulated on Prestel, all information providers are asked to sign a voluntary code of conduct, most of which is concerned with advertising.[175] The basic principle, not surprisingly, is that the advertising "must be legal, decent, honest and truthful."[176] An advertisement must be clearly distinguishable from adjacent data, and viewers must be warned if they are to be charged for a "frame" (video page) of advertising.[177] As to the direct sale of merchandise via Prestel, the Code requires not only delivery within twenty-eight days (or the customer must be given a written notice of a right to cancel the order and obtain a refund), but also a minimum fourteen day customer approval period.[178] If the goods do not correspond to the description or if they are defective, the seller must provide a complete refund.[179]

Some of the more advanced cable systems, which offer interactive capacity and opportunities for direct selling, have much in common with Prestel. Because it is a common carrier medium, however, Prestel's code applies to the information providers, not to the medium of distribution. In its present state, cable appears to be a hybrid between a common carrier (with no control over content) and a publisher (exercising total control over content). If a self-regulatory code were developed for cable, should only the program suppliers (e.g., the cable satellite networks such as CNN and ESPN) be bound? Only the advertisers? Or should the local system operators themselves be responsible for advertising? The cable operators do not act as common carriers, but seek to control the mix of programs offered to subscribers in order to maximize their revenues, thus providing a basis for a share of responsibility for the commercial messages they choose to carry.[180]

As for the national cable programming networks, they are members of NCTA and have representatives on the advertising committee. However, no clearance procedure comparable to the broadcast network practice has yet been established, on the theory that the commercials seen on cable are

---

174. *Oracle Publishes Its First Rate Card*, DIRECT MARKETING, Mar. 1981, at 126.
175. *See* ASSOCIATION OF VIEWDATA INFORMATION PROVIDERS, LTD., VIEWDATA CODE OF PRACTICE (1980).
176. *Id.* § 2, para. 2 (Advertising).
177. *Id.* § 2, para. 3 (Advertising).
178. *Id.* § 3, paras. 3.1 to 3.4 (Direct Sale of Goods and Services).
179. *Id.* § 3, para. 3.7 (Direct Sale of Goods and Services).
180. At least one cable system, Gill Cable of San Jose, California, has a continuity acceptance department and uses the standards set forth in the NAB Code to screen commercials. Statement of Robert Hosfeldt of Gill Cable, The Consumer and Cable Television: A National Conference, Washington, D.C. (Feb. 27, 1981).

no different from those on broadcast television which have already undergone network clearance.[181] If and when the advertising community starts developing campaigns specifically for cable distribution, however, the role of the cable networks in screening commercials could become as crucial as broadcast network clearance is today. If the cable networks do not assume this function, the work of self-regulatory bodies such as the NAD/NARB will certainly be much more difficult.

As may be inferred from the preceding discussion, self-regulation of television advertising has ebbed and flowed over the years, depending on the political climate. Nevertheless, this process has produced benefits to the public in the form of more accurate advertising,[182] which in turn can spur competition in the marketplace. The cable industry has the opportunity to prevent problems from arising that might otherwise require government regulation. In the words of an *Advertising Age* editorial, "it would be foolhardy for business leaders to declare victory and go back to the old ways."[183]

IV. LOCAL CABLE FRANCHISING PROCESS—OPPORTUNITY FOR CITIZENS TO WORK WITH INDUSTRY ON ADVERTISING AND MARKETING ISSUES

A. Local Cable Franchising

Unlike broadcasters, cable operators are not granted licenses by the federal government but rather are franchised at the local level. At one time, however, a cable system could not legally begin providing service without first obtaining a certificate of compliance from the FCC. The prerequisite for obtaining such a certificate was a local franchise granted to the applicant in accordance with a set of procedures and guidelines imposed by the FCC.[184] In the wake of the Supreme Court case overturning some of its major cable television regulations,[185] the FCC deleted the certificate of compliance requirement in favor of the more permissive "signal registration," which need state little more than the applicant's name, address and

---

181. Telephone interview with Kay Koplovitz, President, USA Network (Feb. 13, 1981).
182. *Cf.* Dowling, *Information Content in U.S. and Australian Television Advertising*, 44 J. OF MARKETING 34 (1980), in which the author concluded that Australian television commercials were more informative than their United States counterparts, due in part to the stronger enforcement procedures of Australian self-regulatory authorities. *Id.* at 36-37.
183. *Self Regulation and FTC's New Course*, ADVERTISING AGE, Feb. 9, 1981, at 18.
184. 47 C.F.R. § 76.11 (1972).
185. FCC v. Midwest Video Corp., 440 U.S. 689 (1979).

a skeletal outline of the proposed cable service.[186] Previously mandatory FCC guidelines have been changed to recommendations (e.g., that a full public hearing affording due process precede the grant of a franchise, a fifteen year maximum term for franchise awards, and a requirement that the franchise recipient establish procedures for investigating and resolving subscriber complaints).[187]

The states have also exerted a limited amount of regulatory power over cable television. The Supreme Court has upheld state regulation of cable as a public utility because cable television is local in nature and does not demand national uniformity. The Court further found that the FCC has not preempted the field by regulating in the same subject areas as the states (e.g., rates and quality of signal).[188] Although forty-one states have enacted at least one statute applicable specifically to cable television, only eleven states have adopted any form of cable regulation.[189] Massachusetts,[190] Minnesota,[191] and New York[192] have set up independent cable television commissions, but the actual franchising authority remains with the local governments. Indeed, one of the major functions of the New York State Commission on Cable Television is to advise municipalities on the franchising process.[193] Only in Alaska,[194] Connecticut,[195] Hawaii,[196] Delaware,[197] Rhode Island,[198] and Vermont[199] does the state itself grant

---

186. 47 C.F.R. § 76.12 (1980).
187. 47 C.F.R. § 76.31 (1980). Only the limits on the franchise fee have been retained as mandatory.
188. TV Pix, Inc. v. Taylor, 304 F. Supp. 459 (D. Nev. 1968), *aff'd per curiam*, 396 U.S. 556 (1970) (upholding the constitutionality of a Nevada statute granting regulatory authority over cable television to the Nevada Public Service Commission). More recently, the Court held that municipalities would not be exempt from antitrust actions if local restrictions on competition in the cable area were not specifically authorized by the states. *See* Community Communications Co. v. City of Boulder, 50 U.S.L.W. 4144 (Jan. 13, 1982). *But see Id.* at 4149 (Rehnquist, J., dissenting) (discussing whether federal antitrust laws should preempt local ordinances).
189. FCC, CABLE TELEVISION STATE REGULATION—A SURVEY OF FRANCHISING AND OTHER STATE LAW AND REGULATION ON CABLE TELEVISION 1 (1977). *See generally*, Briley, *State Involvement in Cable TV and Other Communications Services: A Current Review*, in 2 THE CABLE/BROADBAND COMMUNICATIONS BOOK 35 (M. Hollowell ed. 1980).
190. MASS. GEN. LAWS ANN. ch. 166A, §§ 2 & 3 (West 1976 & Supp. 1981).
191. MINN. STAT. ANN. §§ 238.06 & 238.08 (West Supp. 1981).
192. N.Y. EXEC. LAW §§ 814 & 819 (McKinney Supp. 1980).
193. *See* N.Y. STATE COMMISSION ON CABLE TELEVISION, CABLE TELEVISION FRANCHISING WORKBOOK (1980).
194. ALASKA STAT. §§ 42.05.141 & 42.05.701 (1976).
195. CONN. GEN. STAT. ANN. § 16-330 to -333 (West Supp. 1981).
196. HAWAII REV. STAT. §§ 440G-2 to -5 (1976).
197. DEL. CODE ANN. tit. 26, § 201 (Supp. 1980). The Delaware Public Service Commission grants franchises only in unincorporated areas of the state.
198. R.I. GEN. LAWS §§ 391-93 & 391-96 (Supp. 1980).

cable television franchises through the public utility or public service commission.

In most cases, the municipality plays the leading role in regulating cable television. The cable operator generally enters into a franchising agreement with the local government. It is important to realize that a local cable franchise is, above all else, a public franchise, i.e., a legal entity with a long history of judicial interpretation. A cable franchise, like any public franchise, is a privilege conferred by government on an individual or corporation to do for the public benefit that which is not a common right of citizens generally.[200] The laying of cable in, on, and over public streets is a privilege which, but for the grant of a franchise, would be a trespass.[201]

In essence, a cable franchise is a contract between the grantor sovereign, the local government, and the grantee, the cable operator.[202] A franchise offer, even when contained in a statute or ordinance, must be accepted and acted upon to become effective.[203] Once accepted, the franchise is generally treated as any other contract, binding both parties to its terms and tenor.[204] A grantor sovereign can attach to a franchise any terms it sees fit, and may attach any lawful conditions (which then become part of the contract) to its exercise.[205] The only limitations on terms that may be attached to a franchise are that the terms must be legal and not against public policy.[206] A cable franchise may thus include stipulations as to standards of service to the community.[207]

Indeed, a grantor sovereign can, by means of a voluntarily accepted franchise, impose obligations upon a grantee which would be beyond the

---

199. VT. STAT. ANN. tit. 30, §§ 502 & 503 (Supp. 1981).

200. New Orleans Gas Co. v. Louisiana Light Co., 115 U.S. 650, 669 (1885).

201. Cf. People ex rel. Foley v. Begole, 56 P.2d 931 (Colo. 1936); People ex rel. Metropolitan St. Ry. v. State Bd. of Tax Comm'rs, 174 N.Y. 417, 67 N.E. 69 (1903), aff'd, 199 U.S. 1 (1905).

202. See Larson v. South Dakota, 278 U.S. 429 (1929).

203. Capital City Light & Fuel Co. v. Tallahassee, 186 U.S. 401, 409 (1902).

204. Grand Trunk W. Ry. v. South Bend, 227 U.S. 544 (1913); Louisville v. Cumberland Tel. & Tel. Co., 224 U.S. 649 (1912). As a wise precaution, some franchise ordinances specifically state that they are to be treated as contracts. See, e.g., 8 AM. JUR. LEGAL FORMS 2D Franchises from Public Entities § 124:71 (1972).

205. Southern Pac. Co. v. Portland, 227 U.S. 559 (1913).

206. E.g., Southern Pub. Util. Co. v. Charlotte, 179 N.C. 151, 101 S.E. 619 (1919); Chicago Gen. Ry. v. Chicago, 176 Ill. 253, 52 N.E. 880 (1898). They also must not violate the antitrust laws. See Community Communications Co. v. City of Boulder, 50 U.S.L.W. 4144 (Jan. 13, 1982).

207. For sample language stipulating some standards of service for a cable television franchise, see 8 AM. JUR. LEGAL FORMS 2D Franchises from Public Entities § 124:26 (1972) (secs. 1 & 2).

authority of the grantor, but for the grantee's voluntary acceptance.[208] Such obligations attached to a cable television franchise were upheld in *Illinois Broadcasting Co. v. Decatur*:[209]

> Decatur has the right to attach—but not impose or exact—conditions to such use, though unrelated to it, leaving it up to the person seeking the grant to either accept or reject them. Cast in this mold the ordinance becomes effective only upon acceptance. Rejection, and the ordinance is a dead letter. We see nothing wrong in the city saying in effect: You may use our streets for such and such a purpose as such purpose is reasonable and in the public interest yet before we will grant you this, you must in turn agree to certain conditions collateral to the specific use you desire.
>
> . . . .
>
> . . . [T]hese conditions are self-imposed. They are not imposed by Decatur. If General [the cable franchisee] agrees to be so regulated who can complain. If General accepts a condition to pay a given sum who cares, and if General pays, as we assume it will, who is hurt. If General discovers later on that some of the conditions it agreed to are onerous, it can assuage itself with the thought that surcease is but fifteen years away—a condition, too, it agreed to.[210]

Thus, a municipality can place conditions in the franchise agreement, such as provisions for public access and privacy protections for consumers, that, if accepted by the franchisee, would appear to be perfectly legal.

The public franchise differs from ordinary contracts in that the grantor sovereign is not precluded from subsequently enacting legislation under its police power that regulates the exercise of the franchise so long as it does not interfere substantially with the main object of the franchise.[211] A municipality (grantor sovereign) can also specifically reserve the power to alter the franchise by subsequent enactment.[212] Thus, if all the ramifications of the public interest concerning cable television are not apparent at the time the initial franchise is granted, the municipality can retain some degree of flexibility under public franchise law.

---

208. Southern Pac. Co. v. Portland, 227 U.S. 559 (1913).
209. 96 Ill. App. 2d 454, 238 N.E.2d 261 (1968).
210. *Id.* at 461, 238 N.E.2d at 265.
211. Pearsall v. Great N. Ry., 161 U.S. 646 (1896); New Orleans Gas Co. v. Louisiana Light Co., 115 U.S. 650 (1885). The grantor cannot, however, effectively annul the franchise under the rubric of police power. Dobbins v. Los Angeles, 195 U.S. 223, 236 (1904).
212. *Cf.* Fort Smith Light & Traction Co. v. Board of Improvement of Paving, 274 U.S. 387, 390 (1927). For sample language specifically reserving the right to alter the terms of a cable television franchise, see 8 AM. JUR. LEGAL FORMS 2D *Franchises from Public Entities* § 124:41 (1972).

B. Consumer Protection Franchise Provisions

Although it is probably within the authority of every municipality that has granted a cable franchise, there is currently no state law or local regulatory action directed at the responsibility of cable franchisees for advertising or marketing originating on cable channels. A few simple paragraphs in the franchise agreement could establish this responsibility, while retaining sufficient flexibility to account for future developments in the form of advertising and marketing on cable, as well as developments in self-regulation by the industry.

Just as the FCC's rather general requirements that local broadcast stations assume responsibility for screening out deceptive advertisements may in part have led to the protective net of self-regulation of advertising that exists in the broadcast community, so too the presence of local provisions of this type could spur the development of clearance procedures by the satellite-distributed cable networks. A local cable system that risks possible litigation for deceptive commercials would no doubt refuse to distribute advertiser-supported cable programs that did not feature some sort of screening process for the featured commercials. Thus, a few well-placed local restrictions could lead to a healthy level of self-regulation on a national scale by cable program suppliers. Such measures could benefit the industry by increasing the credibility of cable television as an advertising medium.

Many franchise agreements already require the operator to establish a local office to handle complaints about reception, installation, or billing.[213] Should the cable company also be asked to use its local office to resolve consumer complaints about late shipment, defective products, or improper billing of products directly marketed on cable? Should a complaint handling service for advertised items be required only if the cable company does the billing for the marketer or receives a share of the revenue of items sold? Could complaint handling be done more efficiently by pooling the resources of cable systems at the regional level?[214] Given that consumers are being asked to purchase items sight unseen, should the cable system require marketers who use their facilities to offer "satisfaction guaranteed"

---

213. The FCC now recommends establishment of such an office in the franchise agreement. 47 C.F.R. § 76:31 note (1980). For sample franchise language, see 8 AM. JUR. LEGAL FORMS 2D *Franchises from Public Entities* § 124:27 (Supp. 1981) (art. 18).

214. A large portion of advertising on cable systems is sold through regional "interconnects." Gay, *The Cable Rep Entrepreneurs*, MARKETING AND MEDIA DECISIONS, Feb. 1981, at 72.

or at least provide a warranty against defects?[215] At the very least, should the cable system agree to keep records of the addresses and telephone numbers of companies engaging in direct marketing, so that subscribers will have a place to go with their complaints?

Given the somewhat uncertain aspects of the details of advertising and marketing on cable television, the franchise agreement itself could contain only a general provision regarding the cable operator's duties, with the details to be developed over time by a citizens' advisory committee or the local office of telecommunications. In this way, if self-regulation is working, there may be no need for more specific provisions. But it is important at least to establish the basic principle of operator responsibility in the franchise agreement because there may not be an equivalent opportunity to raise the issue until the franchise is renewed some ten or fifteen years later. Once the franchise has been granted, the local community may not have the same bargaining power to win concessions from the cable company. Other issues could also be discussed with franchise applicants. In the interest of assuring the free flow of nondeceptive commercial speech, cable systems could be advised by the local government not to place any unnecessary categorical restrictions on comparative, professional, or price advertisements. The Supreme Court has struck down such advertising restrictions when imposed by the state itself or by professional associations.[216]

Freely-negotiated prophylactic measures could prevent the use of cable television for fraud or deception by inspiring the development of a self-regulatory framework, with the least amount of government involvement. The following language, while by no means foolproof, might be considered to achieve this end:

> FRANCHISEE assumes ultimate responsibility for eliminating any false, deceptive or misleading commercial material from all origination cablecasting material carried on the CABLE SYSTEM.[217]

---

215. Marketers who use the British Prestel system have agreed to sell items "on approval" only. See *supra* note 178 and accompanying text.

216. Bates v. State Bar of Ariz., 433 U.S. 350 (1977) (holding a bar association ban on attorney advertising unconstitutional); Virginia State Bd. of Pharmacy v. Virginia Citizens Consumer Council, Inc., 425 U.S. 748 (1976) (striking down a state ban on price advertising of prescription drugs on first amendment grounds).

217. FCC defines origination cablecasting as: "Programming (exclusive of broadcast signals) carried on a cable television system over one or more channels and subject to the exclusive control of the cable operator." 47 C.F.R. § 76.5(w) (1980). Presumably, the franchise agreement would define it similarly. This definition would not include leased or public access channels.

FRANCHISEE'S responsibility is personal and may not be delegated, but FRANCHISEE may rely on prior determinations regarding false, misleading or deceptive content made by responsible cable television industry networks or organizations, responsible federal regulatory agencies or [state and/or local equivalent agencies].

FRANCHISEE shall provide facilities and personnel for receiving complaints from SUBSCRIBERS regarding products and services advertised over the CABLE SYSTEM. FRANCHISEE shall provide a complaining SUBSCRIBER with the true business name, address and telephone number of the product or service provider and, if requested by the SUBSCRIBER, refer the complaint to an appropriate consumer aid organization. In addition, FRANCHISEE shall compile and retain for three years a log of all such SUBSCRIBER complaints and shall consider any such complaints in determining the acceptability of specific advertisements.

Clearly, each locality would have to tailor the franchise language to suit its needs. One would not want to ask the cable company to produce more consumer protection than is realistic. After all, some systems envision over 100 channels of programming. Broadcast licensees must keep track of the commercials on only one channel. Lack of cable operator control over leased channels and public access channels suggests that these should be excluded from the cable operator's responsibility. Yet, cable systems can be selective about the programs they choose to fill their origination channels. They could demand that national cable networks screen their commercials for possible deception. They could refuse to accept direct marketers for local spots if they do not offer suitable warranties, or if they have been subject to a high level of complaints concerning late shipment. While cable companies may take on this task voluntarily, given the economic realities of the marketplace, it is unlikely that they would do so unless it proved necessary to secure the franchise. In the long run, however, responsible business practices may lead to a greater consumer acceptance of marketing via cable television as well as greater profits for the industry.

C. First Amendment Issues

Some local governments may hesitate to seek consumer protection provisions because they might be challenged on constitutional grounds.[218]

---

218. *See Cable More Akin to Newspapers than Broadcasters, NCTA Tells Senate*, BROADCASTING, May 4, 1981, at 71, which quotes the NCTA: "Like newspaper publishers, cable

132 / Creating Original Programming for Cable TV

Such fears are largely unfounded, however, due to the unique nature of the cable medium and the quasi-contractual nature of the franchising process.

The first amendment protection of the press varies with the medium being used. The Supreme Court has stated that while "the basic principles of freedom of speech and the press . . . do not vary," any one medium is not "necessarily subject to the precise rules governing any other particular method of expression. Each method tends to present its own peculiar problems."[219] The Court has also asserted that "differences in the characteristics of new media justify differences in the First Amendment standards applied to them."[220]

The print medium enjoys the highest level of free press protection. The constitutional doctrine prohibiting prior restraints originated with the written press[221] and is still most forceful in that medium.[222] While licensing is commonly accepted for broadcasting, the licensing of a newspaper would be a classic example of an impermissible prior restraint.[223] A newspaper may need certain licenses as adjuncts to its commercial operation (e.g., building permits and vendors' licenses) but cannot be saddled with a license requirement simply to publish, as a cable or broadcast system is to transmit. Even adjunct licenses, if used to inhibit newspapers, may be impermissible restraints.[224] In fact, the only major restriction on the free exercise of expression placed on the print media is a limited liability for defamation.[225]

With regard to commercial advertising, some limited government restrictions have been permitted. For instance, a paper may be required to remove sex designations from its "help wanted" classified advertisements.[226] In addition, many states have statutes (albeit, unenforced) which hold the publisher responsible for knowingly disseminating false or fraud-

---

operators should be free of government attempts to tell them what must be said on, or who must have access to, their medium of expression."

219. Joseph Burstyn, Inc. v. Wilson, 343 U.S. 495, 503 (1952).
220. Red Lion Broadcasting Co. v. FCC, 395 U.S. 367, 386 (1969).
221. Near v. Minnesota, 283 U.S. 697 (1931).
222. *See* New York Times Co. v. United States, 403 U.S. 713 (1971) (U.S. government could not enjoin the New York Times and the Washington Post from publishing a classified report officially entitled "History of U.S. Decision-Making Process on Viet Nam Policy" ("The Pentagon Papers")).
223. Near v. Minnesota, 283 U.S. at 713.
224. Grosjean v. American Press Co., 297 U.S. 233 (1936).
225. *See* Gertz v. Robert Welch, Inc., 418 U.S. 323 (1974); New York Times Co. v. Sullivan, 376 U.S. 254 (1964).
226. *See* Pittsburgh Press Co. v. Pittsburgh Comm'n on Human Relations, 413 U.S. 376 (1973) (municipal ordinance prohibiting newspapers from carrying sex designations in advertisements for nonexempt job categories did not violate the publisher's first amendment rights).

ulent advertising.[227]

By contrast, a significant amount of regulation is permitted for the broadcast media. While anyone is free to print and distribute written materials, only a small fraction of those who would like to use broadcast frequencies are able to do so without unduly interfering with the broadcasts of others. To deal with this problem of electronic cacophony, Congress established the Federal Radio Commission in 1927.[228] The Communications Act of 1934 created the Federal Communications Commission, giving it the power to allocate frequencies to applicants to serve "the public interest, convenience, or necessity."[229] The scarcity of broadcast frequencies was identified by the Supreme Court as the basis for requiring access for competing viewpoints (i.e., the "fairness doctrine"),[230] a restriction on the free expression of broadcasters that would be impermissible if applied to print publishers.[231]

The Supreme Court has also cited the intrusiveness of the electronic media as a relevant distinction between print publishing and broadcasting for first amendment analysis. Noting that "the broadcast media have established a uniquely pervasive presence in the lives of all Americans" and are "uniquely accessible to children, even those too young to read," the Court upheld an FCC restriction against the broadcast of "indecent" material.[232] In another case, the Court quoted Herbert Hoover's statement from his days as Commerce Secretary that "the radio listener does not have the same option that the reader of publications has—to ignore advertising in which he is not interested . . . ."[233]

Cable television is not identical to print publishing for first amendment purposes but instead has much in common with broadcasting. Anyone can hand out leaflets or establish a newspaper or magazine without interfering with the rights of others. On the other hand, before one can become a cablecaster, a franchise to use the local rights-of-way must be obtained.

Cable television is a video medium equally as intrusive as broadcast television. Both provide sound and moving pictures and have an identical impact on the viewer. Indeed, when a cable system is hooked up to a tele-

---

227. *See Developments in the Law—Deceptive Advertising*, 80 HARV. L. REV. 1122-23, 1152 (1967).
228. Radio Act of 1927, ch. 169, 44 Stat. 1162 (1927) (repealed 1934).
229. Communications Act of 1934, 47 U.S.C. §§ 151-609 (1976).
230. Red Lion Broadcasting Co. v. FCC, 395 U.S. 367, 375-79 (1969); National Broadcasting Co. v. United States, 319 U.S. 190, 216 (1943).
231. Miami Herald Publishing Co. v. Tornillo, 418 U.S. 241 (1974).
232. FCC v. Pacifica Found., 438 U.S. 726, 748-49 (1978).
233. Columbia Broadcasting Sys. v. Democratic Nat'l Comm., 412 U.S. 94, 128 (1973) (upholding the right of broadcasters to refuse to accept paid editorial advertising).

vision set, the typical subscriber foregoes all broadcast television and uses the cable for both cable-originated programming and retransmission of broadcast-originated programming. As far as the impact on the viewer is concerned, the only significant difference between broadcast and cable television is that the former is free, while the latter is only available through paid subscription.

On the other hand, cable television lacks the same scarcity of available frequencies as broadcasting. Current technology allows over fifty channels to be transmitted over a single coaxial cable and, theoretically, any number of cables can be installed at the same time. It is precisely this lack of scarcity of outlets that is most commonly cited for the proposition that first amendment principles applicable to cable television should be those now applied to newspapers, which also lack scarcity, rather than those principles applied to the broadcast media.[234]

While cable does not suffer from the same scarcity of frequencies problem that over-the-air broadcasting does, it is a limited access medium in other ways. Many cable systems derive the bulk of their programming from satellite-distributed networks. The satellites themselves have limited available space, and the launching of communication satellites must be authorized by the FCC.[235] Thus, the distribution of cable programs is limited by the availability of satellite space.

In addition, the capital expenditures necessary to wire a city or town are such that only one cable system is likely to be available in any given locality.[236] While exclusive franchises are rare and have been challenged legally,[237] in effect most cable franchise holders have a regional monopoly. Although the Supreme Court has never ruled expressly on "economic scarcity" (when the relevant market will not support more commercial users, although the medium could physically withstand more), at least one court has inferred from Supreme Court opinions that economic scarcity is not sufficient to justify government regulation of the press.[238] In this sense, the

---

234. *E.g.*, Note, *Cable Television and Content Regulation: The FCC, the First Amendment and the Electronic Newspaper*, 51 N.Y.U.L. REV. 133, 135 (1976).
235. *See generally* 47 C.F.R. § 25.390 (1980).
236. An attorney with the National Cable Television Association has stated: "In reality, if one operator has 'built' a city, another cable operator isn't going to build another system." Andrew, *Courts Ponder Status of Cable TV to Rule on Legality of Regulation*, Wall St. J., Dec. 29, 1980, at 11, col. 3 (quoting James Ewalt).
237. *See California Judge Says Exclusive Franchises Are Unconstitutional*, BROADCASTING, Jan. 12, 1981, at 69. Governor Carey of New York has also proposed legislation aimed at encouraging new cable companies to compete in areas where only one company is now providing service. *Carey Moves to Ease Controls on Cable TV*, N.Y. Times, Apr. 13, 1981, at 81.
238. Home Box Office, Inc. v. FCC, 567 F.2d 9 (D.C. Cir. 1977). That court interpreted

one-cable town is the same as the one-newspaper town. Any first amendment analysis of the permissible limits of government regulation must account not only for the nature of the medium but also for the nature of the proposed restriction itself.[239] The concept of cablecaster responsibility for false or misleading advertising on the channels it controls is far less of an intrusion on editorial prerogatives than are requirements to set aside leased channels or to provide access to any and all speakers on a first come—first served basis. The proposal discussed in this article would affect only commercial speech, which is entitled to a lesser degree of protection than other forms of speech.[240] Furthermore, the only type of speech that might be suppressed, namely false or misleading advertising, is wholly outside the first amendment's protection.[241]

The constitutional status of commercial speech has undergone dramatic change in recent years. The issue had remained dormant since the 1940's, when *Valentine v. Chrestensen*[242] suggested to nearly all courts and commentators that "purely commercial advertising," whether deceptive or not, was wholly without first amendment protection and could, therefore, be freely regulated by the government.

Miami Herald Publishing Co. v. Tornillo, 418 U.S. 241, 247-56 (1974), as indicating that economic scarcity is "insufficient to justify even limited government intrusion into the First Amendment rights of the conventional press . . . and there is nothing in the record before us to suggest a constitutional distinction between cable television and newspapers on this point." 567 F.2d at 46.

239. *Cf.* Home Box Office, Inc. v. FCC, 567 F.2d at 46. ("The absence in cable television of the physical restraints of the electromagnetic spectrum does not, however, automatically lead to the conclusion that no regulation of cable television is valid. . . . [R]ules restricting speech do not necessarily abridge freedom of speech.").

240. Ohralik v. Ohio State Bar Ass'n, 436 U.S. 447, 456 (1978). Several commentators have been critical of the Court's developing commercial speech doctrine, arguing that the balancing approach is a return to "economic due process" because it fails to give appropriate deference to state regulation of commercial activities. *E.g.*, Jackson & Jeffries, *Commercial Speech: Economic Due Process and the First Amendment*, 65 VA. L. REV. 1 (1979). *See also* Central Hudson Gas & Elec. Corp. v. Public Serv. Comm'n, 447 U.S. 557, 588 (1980) (Rehnquist, J., dissenting). Others have criticized the extension of a lesser form of first amendment protection to commercial speech because it may dilute the level of protection afforded other forms of speech. *E.g.*, Baker, *Commercial Speech: A Problem in the Theory of Freedom*, 62 IOWA L. REV. 1 (1976); Roberts, *Toward A General Theory of Commercial Speech and the First Amendment*, 40 OHIO ST. L.J. 115 (1979). Another commentator fears that the application of a first amendment balancing test commercial speech may undercut the general free speech principle of content neutrality. Farber, *Commercial Speech and First Amendment Theory*, 74 Nw. U.L. REV. 372 (1979).

241. Central Hudson Gas & Elec. Corp. v. Public Serv. Comm'n, 447 U.S. at 566 ("At the outset, we must determine whether the expression is protected by the First Amendment. For commercial speech to come within that provision, it at least must concern lawful activity and not be misleading.").

242. 316 U.S. 52, 54 (1942).

The Supreme Court, in the landmark case of *Virginia State Board of Pharmacy v. Virginia Citizens Consumer Council, Inc.*,[243] first held unequivocally that speech that "does no more than propose a commercial transaction" is entitled to some degree of first amendment free speech protection. Recognizing the interests of sellers, consumers, and society, the Court emphasized that the "free flow of commercial information . . . is indispensable to the proper allocation of resources in a free enterprise system . . . ."[244] Yet, from the outset, the Court recognized that there were limits to the first amendment shield in the commercial area. The Court alluded to the "commonsense differences" between commercial speech and other varieties, namely, that its truth is more readily verified by the speaker than other forms of speech, and that it is more durable and less likely to be chilled by regulation because "advertising is the *sine qua non* of commercial profits . . . ."[245] Acknowledging the legitimate government interest in regulating false, deceptive, or misleading advertising, the Court stated that "[t]he First Amendment, as we construe it today, does not prohibit the State from insuring that the stream of commercial information flow cleanly as well as freely."[246]

Given the Court's characterization of commercial speech, it is not surprising that the permissible scope of regulation goes beyond the restriction of misleading or deceptive speech to include broad, prophylactic measures designed to prevent deception.[247] Thus, the Court upheld a disciplinary action against an Ohio lawyer for violating a rule against in-person solicitation, even though it was not alleged specifically that his conduct was deceptive, because "the State has a legitimate and indeed 'compelling' interest in preventing those aspects of solicitation that involve fraud, undue influence, intimidation, overreaching, and other forms of 'vexatious conduct.'"[248] Similarly, the Court upheld a complete ban on the use of trade names by optometrists because it concluded there was "a significant possibility that trade names will be used to mislead the public."[249] In other

---

243. 425 U.S. 748, 762 (1976).
244. *Id.* at 765. For a discussion of the relationship of commercial speech to the marketplace of ideas theory of free speech, see Westen, *The First Amendment: Barrier or Impetus to FTC Advertising Remedies?*, 46 BROOKLYN L. REV. 487 (1980).
245. 425 U.S. at 771-72 n.24.
246. *Id.* at 771-72.
247. For an economic and legal analysis of the appropriate scope of government regulation to prevent deception in commercial speech, see Reich, *Preventing Deception in Commercial Speech*, 54 N.Y.U.L. REV. 775 (1979).
248. Ohralik v. Ohio State Bar Ass'n, 436 U.S. 447, 462 (1980).
249. Friedman v. Rogers, 440 U.S. 1, 13 (1979). Some authors have characterized the *Friedman* opinion as a dramatic shift away from the more demanding balancing test applied to strike down state laws in *Virginia Pharmacy* and *Bates. E.g.*, Comment, *First Amendment*

words, the "government may ban forms of communication more likely to deceive the public than to inform it . . . ."[250]

The Court has recently clarified the test for government regulation of commercial speech in *Central Hudson Gas & Electric Corp. v. Public Service Commission*,[251] a case involving a state regulation prohibiting a utility from advertising to promote the use of electricity. The majority opinion stated that a "four-part analysis" has developed for commercial speech cases. First, a court must determine whether the expression comes within the first amendment's purview at all. For commercial speech, that means commercial speech "at least must concern lawful activity and not be misleading."[252] Second, the government must assert a "substantial interest" to be accomplished by the regulation.[253] Third, the restriction must "directly advance" the substantial state interest.[254] Fourth, the reviewing court must determine whether the restriction "is not more extensive than is necessary."[255] For the purpose of local regulation of false or misleading commercial speech on cable television, stopping at the first inquiry is appropriate. Such commercial speech is utterly without first amendment protection and may be prohibited by government regulation.

The Supreme Court has not yet dealt with the issue of the legal responsibility of the *publisher* of deceptive advertising, as opposed to the responsibility of the *advertiser* himself. By analogy, however, cases holding publishers liable for defamation do not distinguish between liability for material originating with the publisher and material submitted by others, including advertisements.[256] Clearly publishers, and even broadcasters,

---

Protection for Commercial Speech: An Optical Illusion?, 31 U. FLA. L. REV. 799 (1979); Note, *Reuniting Commercial Speech and Due Process Analysis: The Standard for Deceptiveness in Friedman v. Rogers*, 57 TEX. L. REV. 1456 (1979).

250. Central Hudson Gas & Elec. Corp. v. Public Serv. Comm'n, 447 U.S. at 563.
251. 477 U.S. 557 (1980).
252. *Id.* at 566.
253. *Id.* at 564.
254. *Id.* at 566.
255. *Id.* The New York regulation banning all promotional advertising by electric utilities was held invalid because the Public Service Commission failed to show that less restrictive methods would not achieve its admittedly substantial interest in energy conservation. It is not clear that this "least restrictive means" test would apply to restrictions aimed at misleading or deceptive speech because such expression is apparently considered by the Court to be outside the scope of the first amendment. Also, in an earlier case in which a state banned all trade names for optometrists because they were potentially deceptive, the Court said that "there is no First Amendment rule . . . requiring a State to allow deceptive or misleading commercial speech whenever the publication of additional information can clarify or offset the effects of the spurious communication." Friedman v. Rogers, 440 U.S. 1, 12 n.11 (1978).

256. *See* Louisville Times Co. v. Lyttle, 257 Ky. 132, 77 S.W.2d 432 (1934); Kulesza v. Alliance Printers & Publishers Inc., 318 Ill. App. 231, 47 N.E.2d 547 (1943); Fitch v. Daily

have the freedom to accept or reject paid commercial messages,[257] so that some reasonable duty of care with regard to false or deceptive advertisements is well within the accepted scope of editorial control. This responsibility should not have a chilling effect on the dissemination of truthful commercial speech on cable television, nor should it unduly burden the protected journalistic functions of disseminating news, information and entertainment, because the cablecaster as well as the advertiser stands to gain revenues through the distribution of advertising.

The conclusion that cablecaster responsibility for false or misleading advertising would be constitutional becomes even more forceful when it is recalled that the municipality would not be simply regulating the local cable system, but would *in effect* be conditioning a privilege on the acceptance of some degree of responsibility for its use. It is beyond question that a municipality could prohibit altogether a private company from wiring its streets unless state law dictated otherwise. Since such wiring is not a right, a municipality could reasonably condition the grant of the privilege on the acceptance of certain conditions. The conditions which may be imposed, however, are not without limit. The Supreme Court has stated:

> [A]s a general rule, the state, having power to deny a privilege altogether, may grant it upon such conditions as it sees fit to impose. But the power of the state in that respect is not unlimited; and one of the limitations is that it may not impose conditions which require the relinquishment of constitutional rights.[258]

Since it is difficult to argue that the dissemination of deceptive commercial speech is a constitutional right, however, this doctrine should not pose a significant problem.

The Supreme Court has held that surrender of substantial first amendment rights (indeed, the imposition of a clear prior restraint) is not a violation of public policy when such a surrender is part of a voluntary agreement. This would be the case in a cable franchise agreement. In the

---

News Pub. Co., 116 Neb. 474, 217 N.W. 947 (1974). Several FTC cases have included the advertising agency as well as the advertiser in actions against deceptive commercial messages. *See, e.g.*, Sears, Roebuck & Co., 95 F.T.C. 406 (1980). *But see* New York Times Co. v. Sullivan, 376 U.S. 254, 266 (1964), in which the Supreme Court noted that noncommercial speech does not lose its protection (and the publisher cannot be held liable) solely because it is published as a paid "editorial" advertisement.

257. Columbia Broadcasting Sys. v. Democratic Nat'l Comm., 412 U.S. 94 (1973).

258. Frost & Frost Trucking Co. v. Railroad Comm'n of Cal., 271 U.S. 583, 593-94 (1926). The continuing validity of *Frost* has been questioned. Geophysical Corp. of Alaska v. Andrus, 453 F. Supp. 361, 371 n.2 (D. Alaska 1978). Moreover, its literal application appears to contravene the subsequently enacted Communications Act of 1934, 47 U.S.C. §§ 151-609 (1976) (conditioning the privilege of a broadcast license on acceptance of restrictions of first amendment rights in its use).

recent case of *Snepp v. United States*,[259] the Court concluded that the CIA may require, in an employment contract, the submission of any writings (even if conceded to contain no classified material) for prepublication review. Just as a cable applicant can refuse conditions, the prospective employee in the *Snepp* case could have refused to agree to the condition and thus sacrificed government employment.[260] If a voluntary agreement to a prior restraint on free expression is not against public policy, it seems safe to conclude that a voluntary agreement to avoid the dissemination of false or misleading advertising (not protected by the first amendment) would not be proscribed.

In summary, the local franchising process provides an opportunity to negotiate with prospective cable operators regarding the public's interest in preventing the dissemination of deceptive advertisements or abusive marketing schemes on cable television. First amendment considerations, while not to be taken lightly, should not preclude a cable company from agreeing to accept responsibility for commercial messages in return for the grant of the franchise. Only unprotected speech, i.e., false or misleading advertising, would be affected. Further, because commercial speech is readily verifiable and peculiarly hardy, reasonable measures to preclude the use of cable for deceptive advertisements should neither chill the flow of truthful commercials nor reduce the cablecaster's economic base.

V. Conclusion

Cable television is on the verge of maturing as a conduit for commercial messages and original program services. Thus far, no blemish has been discovered to mar cable's image as a bonanza of information and entertainment. Yet some instruction can be gleaned from examining typical consumer problems, such as deceptive advertising and unfair marketing practices, along with the government and industry regulatory framework that has kept such practices in check in the traditional media.

The pressure of FCC regulation and threats of greater government involvement appear to have played a role in the development of self-regulation of advertising in broadcasting. Yet the FCC's jurisdiction to extend similar restrictions to cable television is uncertain. Furthermore, it would be inappropriate to apply industry-wide federal requirements for cable television advertising unless a record of widespread consumer injury had developed. Similarly, the extension of FTC provisions regarding cooling-off

---

259. 444 U.S. 507 (1980).
260. *Cf.*, United States v. Marchetti, 466 F.2d 1309, 1316 (4th Cir.), *cert. denied*, 409 U.S. 1063 (1972).

periods and late shipment for home sales should be approached with caution lest a promising new medium for commerce be strangled by the premature imposition of unnecessary restrictions.

Responsible self-regulation, with input from citizens and government, could be the most cost-effective means of preventing consumer injury as well as maintaining cable television's credibility with its viewers. The local franchising process offers an opportunity for citizens and industry to negotiate acceptable measures to deal with consumer protection issues in a cooperative, rather than an adversarial, setting. This type of low-key, flexible experimentation at the local level could serve to bolster the resolve of the industry to police itself. By acting now, consumer injury could be avoided and the necessity for burdensome litigation and unwieldy adminstrative procedures foregone. As the decade of cable television unfolds, let us hope that a spirit of fair play will prevail.

Glossary

In using this glossary, the reader needs to be aware that occasionally some terms over-lap, occasionally some terms which originally had different meanings are now used interchangeably, and in some cases no one standard term has been formally or informally agreed upon. An attempt has been made to be as clear, accurate, and specific as possible in dealing with these problems. Also, some of the terms included here have other definitions for use in other fields; for the most part, this glossary defines the terms only as they apply to cable/broadband.

Copyright 1980, 1983 by Communications Press, Inc.

access channels: channels which cable operators make available to others. Under rules the FCC enacted in 1972, cable operators in the top 100 TV markets had to offer four categories of access channels: public, educational, government, and leased. Although the U.S. Supreme Court overturned the rules, the terms are still in use. Many cable operators voluntarily offer access channels, and some states and cities require them.

addressability: the capability of sending signals downstream from the headend to specific locations or "addresses" on the cable system. This service requires an addressable converter at the subscriber's terminal.

addressable, programmable converter/descrambler (APCD): refers to a single subscriber terminal for cable TV in which all four functions are combined.

ADI: area of dominant influence. Term used by the Arbitron ratings service to indicate the area in which a single television station can effectively deliver an advertiser's message to the majority of homes. (Nielsen's term is DMA [designated market area].)

aerial plant: cable installed on a pole line, or similar overhead structure; space often leased from the local telephone or power company. Compare with *buried plant.*

amplifier: device used to increase the power, voltage, or current of input signal. Used in a cable system's distribution plant to overcome the effects of attenuation caused by the coaxial cable.

amplitude: magnitude or range of voltage, power, etc. of an electrical signal or radio wave.

analog: transmission format that is a continuously variable signal, generally represented by a flowing wave. Compare with *digital.* To convert signals from one

144 / Glossary

of these transmission formats to the other, some form of modem (modulator/demodulator) is used.

antenna: device used to aid in the reception or transmission of signals.

APCD: *addressable, programmable converter/descrambler.*

ascertainment: in broadcasting, a requirement that stations determine the needs and interests of the community, through interviews and surveys, and design programming to help meet those needs and interests. Ascertainment studies are sometimes used by cable television companies as well.

aspect ratio: ratio of width to height in television picture.

attenuation: a decrease in the amplitude of a signal as it progresses from source to receiver, such as occurs in a coaxial cable. See *amplifier.*

automated channel: a channel on a cable system dedicated for information provided in an alphanumeric or graphic format utilizing a character generator. Some examples of automated channels are time/weather, news updates, program guides, swap and shop, community calendar, travel and entertainment, financial news, consumer assistance, and real estate listings.

bandwidth: measure of the information-carrying capacity of a communication channel. The bandwidth corresponds to the highest frequency signal which can be carried by the channel.

baseband: audio and video signal prior to its conversion to a frequency more useful for transmission via a particular medium.

basic cable service: the service that cable subscribers receive for the threshold fee—including local television stations, some distant signals, and a number of non-broadcast signals, depending on the channel capacity of the system.

bird: *satellite.*

branch cable: a secondary section of cable leading from the trunk, or distribution cable, past subscribers' homes. Also known as feeder cable.

bridging: a cable television amplifier which takes a small amount of signal from the trunk, amplifies it, and feeds it to one or more feeder lines.

broadband: relative term referring to a system which carries a wide frequency range (sometimes used to refer to frequency bandwidth greater than one MHz). In a telephone-television context, telephone would be considered narrowband (3kHz) and television would be considered broadband (6 MHz). *Broadband* is also often used to refer to communications systems such as cable TV, satellite, and microwave relay systems, which carry a large number of simultaneous transmissions. Also known as *wideband.*

broadcaster's service area: the geographical area covered by a broadcast station's signal. Also see *Grade A contour* and *Grade B contour.*

broadcasting: transmission of information by electromagnetic means, intended for public reception. Compare with *narrowcasting.*

broadcasting-satellite service: defined by the 1971 World Administrative Radio Conference for Space Telecommunications (1971 WARC-ST) as a radio-communication service in which signals transmitted or retransmitted by space stations are intended for direct reception by the general public. In an appending note the definition provided that "direct reception" encompasses both individual reception and community reception. These modes were defined as follows:

individual reception: the reception of emissions from a space station in the broadcasting-satellite service by simple domestic installations and in particular those possessing small antennae.

community reception: the reception of emissions from a space station in the broadcasting-satellite service by receiving equipment, which in some cases may be complex and have antennae larger than those used for individual reception, and intended for use: by a group of the general public at one location; or through a distribution system covering a limited area.

Compare with *fixed-satellite service,* and see also *direct broadcast satellite.*

BSS: *broadcasting-satellite service.*

buried plant: cable installed under the surface of the ground. Compare with *aerial plant.*

cable: transmission medium designed to carry either electronic or digital information over conductive or optical lines.

cable radio: broadcast and non-broadcast audio programming distributed by cable TV systems. The service predominates in areas where few FM radio signals are available over-the-air.

"cable-ready" television set: an "improved" TV set which has adequate shielding to allow use in strong local fields, and whose tuner will tune cable channels as well as standard over-the-air broadcast channels. In cable systems offering multiple pay channels, where the subscriber may choose one or more and may start and stop any of them whenever he wishes, the so-called cable-ready TV set is no longer cable-ready. The newest means of handling such services is the addressable programmable converter/descrambler (APCD). (A nationally standardized system of coding and addressing could allow these functions to be built into the subscriber's set along with cable-ready tuning.)

cable television channel: a transmission path which is used by a cable television system to deliver a signal to or from subscribers. See *Class I, II, III,* and *IV cable television channels.*

cable television system: a broadband communications system, capable of delivering multiple channels of entertainment programming and non-entertainment information, generally by coaxial cable. Many cable TV designs integrate microwave and satellite links into their overall design, and some now include optical fibers as well. Technically, for FCC purposes, a cable television system is "a non-broadcast facility consisting of a set of transmission paths and associated signal generation, reception, and control equipment, under common ownership and control, that distributes or is designed to distribute to subscribers the signals of one or more television broadcast stations, but such term shall not include (1) any such facility that serves fewer than 50 subscribers, or (2) any such facility that serves or will serve only subscribers in one or more multiple dwellings under common ownership, control, or management." Sometimes called cable system or *CATV.*

cablecasting: programming other than broadcast signals carried on cable television systems. See *Class I, II, III,* and *IV cable television channels.* Also see *local programming, local origination, origination cablecasting,* and *access channels.*

carrier: an electromagnetic wave some characteristic of which is varied in order to convey information, and which is transmitted at a specified frequency. For example, a television signal includes a video carrier and an audio carrier within the 6 MHz channel bandwidth.

CARS: *Community Antenna Relay Service.*

cascade: the operation of two or more devices (such as amplifiers in a cable television system) in sequence, in which the output of one device feeds the input of the next, thus allowing equally strong signals to reach all portions of the system.

CATV: Community Antenna Television, a term used to refer to cable television. See *cable television system.*

Certificate of Compliance: under an FCC regulatory scheme in effect from 1972 to 1978, cable system operators had to apply for and receive Certificates of Compliance prior to the commencement of service or the addition of signals. Since elimination of the requirement in 1978, cable operators need file only Registration Statements when they commence operations or add signals, and need not await further FCC action.

channel: a signal path for conveying information.

channel capacity: in a cable TV system, the number of channels that can be simultaneously carried on the system. Generally defined in terms of 6 MHz (television bandwidth) channels.

character generator: electronic device which generates letters, numbers, or symbols, directly on a television screen. Usually used to supply paginated information, such as news, time, weather, etc., for display on subscribers' TV receivers.

churn: rate of turnover in subscriptions to cable and/or pay television service.

Class I cable television channel: as defined by the FCC, a signalling path provided by a cable television system to relay to subscriber terminals television broadcast programs that are received off-the-air or are obtained by microwave or by direct connection to a television broadcast station.

Class II cable television channel: as defined by the FCC, a signalling path provided by a cable television system to deliver to subscriber terminals television signals that are intended for reception by a television broadcast receiver without the use of an auxiliary decoding device and which signals are not involved in a broadcast transmission path.

Class III cable television channel: as defined by the FCC, a signalling path provided by a cable television system to deliver to subscriber terminals signals that are intended for reception by equipment other than a television broadcast receiver or by a television broadcast receiver only when used with auxiliary decoding equipment.

Class IV cable television channel: as defined by the FCC, a signalling path provided by a cable television system to transmit signals of any type from a subscriber terminal to another point in the cable television system.

closed circuit: any communication transmission method by which reception is not available to the general public. The receiving equipment is directly linked to the originating equipment by cable, microwave relay, satellite, or telephone lines. Extensively used for monitoring in hospitals, police stations, prisons, industrial training and education programs, schools, etc; for long-distance programming to specialized audiences, such as for certain sporting events and political fundraisers; and for teleconferencing.

coaxial cable: type of cable in which the center conductor is surrounded by an insulator in turn surrounded by an outer conductor. Various types of coaxial cable are the most common transmission media in cable television systems.

common carrier: a company which holds itself out to the public as providing a communications service for hire. Included in the definition are companies which own communications satellites, point-to-point microwave facilities, and MDS, as well as telephone and telegraph companies. Some have argued that cable television should be deemed a common carrier, at least insofar as the cable company leases channels.

Communications Act of 1934: federal statute which replaced the Radio Act of 1927 and established the FCC. The purpose of the Act is to "make available, so far as possible, to all the people of the United States, a rapid, efficient, nationwide, and world-wide wire and radio communications service."

Community Antenna Relay Service (CARS): the microwave radio relay service reserved by the FCC for transmitting programming from one point to another by cable TV operators.

compulsory license: legal requirement, under the Copyright Law, for copyright holder to license users of their copyrighted material on a uniform basis and for a stipulated fee. Under this system, cable operators make payment for certain of the broadcast programming they carry, to the Copyright Office, which through its Copyright Royalty Tribunal distributes the payment to copyright holders.

conductor: any material, such as wire or coaxial cable, capable of carrying an electric current.

conduit: thin metal or plastic pipe used for protecting wire or cable.

connector: mechanical or electrical device that is used to join two or more wires, cables, circuits, or components.

contour: predicted, theoretical coverage area of a broadcast station. Under the FCC rules, TV stations have three contours: City Grade, Grade A, and Grade B— each respectively covering a wider area, but each of lesser signal strength.

converter: device for changing the frequency of a television signal. A cable headend converter changes signals from frequencies at which they are broadcast

148 / Glossary

to clear channels (and from UHF to VHF since cable cannot carry UHF signals). A subscriber converter ("set-top converter") extends the channel capacity of the home television receiver. The set-top converter will have either buttons or dial, for the subscriber to select channels.

cross-modulation: see *intermodulation distortion.*

cross-over: spectrum between that used for upstream transmission and that used for downstream transmission in a two-way cable system. Also see *sub band.*

cycle: one complete alternation of a sound or radio wave. The rate of repetition of cycles is the frequency. Also see *hertz.*

DBS: *direct-broadcast satellite.*

decoder: electronic device which translates signals in such a way as to recover the original message or signal. Compare with *encoder.*

dedicated channel: a cable television channel solely used for a particular type of service, such as education, police, meter reading, public access, library, business data services, etc.

digital: a transmission format that involves pulses of discrete or discontinuous signals. Compare with *analog.* To convert signals from one of these transmission formats to the other, some form of modem (modulator/demodulator) is used.

direct broadcast satellite: satellite designed to transmit signals intended for direct reception by the general public. See *broadcasting-satellite service.*

discrete address: transmission from one single point to another single point. Compare with *multiple address.* And see *point-to-point service.*

dish: *earth station.*

display: the visual information shown on a television receiver.

distant signal: signal of a television station which would not be deemed "local" to a given cable television system under FCC's rules. Generally, signal originating at a point too distant to be picked up by ordinary television reception equipment.

distribution system: collective term for the part of the cable television system used to carry signals from the headend to subscriber TV receivers.

DMA: designated market area. Term used by the Nielsen ratings service to indicate the area in which a single television station can effectively deliver an advertiser's message to the majority of homes. (Arbitron's term is ADI [area of dominant influence].)

domsat: abbreviation for domestic (U.S.) satellite.

downlink: satellite-to-ground transmission. Compare with *uplink.*

downstream: in a cable system the direction from the headend to the subscriber terminals. Compare with *uplink.*

drop cable: that cable which feeds the signal to an individual customer from the feeder cable serving the specific area.

Copyright 1980, 1983 by Communications Press, Inc.

dual trunk capability: see *multiple cable system.*

earth station: dish-shaped antenna used for the reception (and if designed for it, transmission) of satellite signals. Sometimes called *dish.*

education access channel: a channel on a cable system dedicated for use by educational entities, generally offered free of charge by the cable system. Also see *access channels.*

electromagnectic spectrum: the continuum of frequencies generally useful for transmission of information or power by electromagnetic means. Differing bands of frequencies are allocated to different types of communications services.

encoder: electronic device that breaks up a signal into component parts for transmission of that signal. Compare with *decoder.*

exclusivity: the sole right to air a program within a given period of time in a given market.

facsimile (FAX, FX): system of telecommunications for the transmission of printed or graphic material, converted into electronic signals at the source and carried to the subscriber's receive terminal where it is reconverted into a copy of the original.

feeder cable: a secondary section of cable leading from the trunk, or distribution cable, past subscribers' homes. Also known as branch cable.

fiber optics: see *optical fiber.*

field: one half of a television picture—the odd or even scanning lines. Two fields are interlaced to form one frame or complete picture, and scanning occurs at 60 fields per second. Also see *frame* and *scanning line.*

filter: circuit within a cable distribution system that allows passage of desired channels and blocks others, such as for distribution of different packages of service.

fixed-satellite service: defined by the 1971 World Administrative Radio Conference for Space Telecommunications (1971 WARC-ST) as a radio-communication service between earth stations at specified fixed points when one or more satellites are used, in some cases including satellite-to-satellite links; and for connection between one or more earth stations at specified fixed points and satellites used for a service other than the fixed-satellite service (for example, the broadcasting-satellite service, etc.). Compare with *broadcasting-satellite service.*

footprint: satellite coverage area.

frame: one complete television picture consisting of two fields of interlaced scanning lines. A frame lasts 1/30 of a second.

frame grabber: control device allowing an individual television viewer to stop a single frame of a moving TV picture and hold that frame for as long as the viewer may wish to see it.

franchise: an authorization or license issued by a political subdivision for the operation of a cable television system. The document usually sets out the specific rights and responsibilities of each party to the agreement. See *ordinance.*

150 / Glossary

franchise fee: annual fee collected by the franchising authority from the cable operator, generally based on a percentage of the cable operator's gross revenues, and used to cover such things as cost of use of the public right-of-way, regulatory activities, and other cable-related expenses. FCC regulations limit the allowable franchise fee to three percent of gross revenues, or five percent with special permission.

frequency: the number of complete alternations of a sound or radio wave in a second. See *hertz, kilohertz, megahertz,* and *gigahertz.*

frequency-division multiplex: technique by which two signals are transmitted simultaneously on the same transmission medium; each signal is assigned to a separate and distinct carrier frequency within the medium. Also see *time-division multiplex.*

FSS: *fixed-satellite service.*

geostationary orbit (GSO): a special class of geosynchronous orbit; a circular equatorial orbit in which a satellite appears to remain stationary with respect to the surface of the earth. Note that the acronym *GSO* is used to refer to both geostationary orbit and geosynchronous orbit. Also see *orbital position.*

geosynchronous orbit (GSO): a satellite orbit with a period of exactly one day. Also see *geostationary orbit* and *orbital position.*

geostationary satellite: orbital communications satellite moving at the speed of the earth's rotation, thus apparently stationary as viewed from the earth's surface. Also known as synchronous or geosynchronous satellite.

geosynchronous satellite: see *geostationary satellite.*

gigahertz (GHz): a unit of frequency equivalent to one billion hertz or cycles per second.

Grade A contour: geographical reception area of a broadcast station wherein satisfactory reception is estimated to be available 90 percent of the time at 70 percent of the receiver locations. Part of the broadcaster's service area. Also see *Grade B contour.*

Grade B contour: geographical reception area of a broadcast station wherein satisfactory reception is estimated to be available 90 percent of the time to 50 percent of the receiver locations. Often spoken of as 35 miles, but actually could be far more or less, depending on terrain factors and on antenna efficiencies at the transmitter.

grandfathering: generally applies to situations in which government changes a policy but allows persons doing the thing newly-prohibited to continue doing it.

GSO: acronym used interchangeably to refer to (1) *geosynchronous orbit* and (2) *geostationary orbit.*

hardware: actual physical equipment, such as cameras, recorders, etc., as distinguished from materials like programming. Compare with *software.*

HDTV: *high-definition television.*

headend: control center of a cable television system, where incoming signals are amplified, converted, and combined in a common cable along with any origination cablecasting, for sending out to subscribers.

hertz (Hz): a unit of frequency equivalent to one cycle per second. See *cycle*.

high definition television (HDTV): any of several improved forms of television photography, recording, transmission, and reception. Improvements may include higher resolution, higher sound fidelity, (generally stereo), more faithful color rendition, and a wider aspect ratio.

home information utility: a service which involves the delivery of "alpha-numeric' material (words and numbers)—and sometimes video and audio as well—to a TV screen. It comes in a variety of forms (see *videotex*, for example). By using a device at home, viewers can request information to be displayed on the TV screen. The home device may be as simple as a pad which resembles a hand-held calculator; or it may be a computer-keyboard which permits complicated interactive "conversations" with the system's computer. The data in the system may be as simple as airline timetables, stock market figures, or recipes; or it can be as complicated as financial planning, budgeting, and tax-return calculations.

homeset: in direct broadcast satellite (DBS) service, refers to the equipment necessary to receive signals—consisting of a dish antenna about three feet in diameter mounted on the rooftop of a single family home, and the electronics necessary to convert and descramble the signals from the satellite.

hub-network: a modified tree-network in which signals are transmitted to subordinate distribution points (hubs) from which the signals are further distributed to subscribers. Also see *tree-network*.

independent station: a commercial television broadcast station that is not affiliated with one of the three major commercial television networks.

infomercial: program-length commercial message. Each advertiser could create its own infomercial or the cable operator could develop special shoppers' programs by grouping commercial messages together. And the cable operator could group infomercials around a special theme, similar to a newspaper's "Home" or "Wedding" sections, for example.

institutional network: in a cable system, a dedicated network (sometimes separate, sometimes integrated with the general subscriber network) for use by institutions such as schools, hospitals, government and non-profit agencies, and business. Sometimes called *special service network*.

Instructional Television (ITV): a television system used primarily for formalized instruction.

Instructional Television Fixed Service (ITFS): the frequencies set aside by the FCC for use by educational institutions in relaying ITV programs.

interactive cable system: see *two-way cable.*

152 / Glossary

interconnection: use of microwave, satellite, coaxial cable, optical fiber, or other apparatus or equipment for the transmission and distribution of signals between two or more cable systems for mutual distribution of programming.

interference: extraneous disturbance that causes degradation or disruption of normal signal transmission.

intermodulation distortion: form of interference involving the generation of interferring frequencies during signal processing.

ITFS: *Instructional Television Fixed Service.*

ITV: *Instructional Television.*

kilohertz(kHz): a unit of frequency equivalent to one thousand hertz or cycles per second.

laser: from Light Amplification by Stimulated Emission of Radiation. Device for transmitting light in coherent form.

leased access: a channel on a cable system dedicated for lease to seaprate program providers; usually utilized by for-profit entities. Also see *access channels.*

local distribution service (LDS): a communications system used for point-to-multipointservice in a local area. The technology used may be telephone lines, microwave, or coaxial cable.

local government access channel: a channel on a cable system dedicated for use by local government and municipal entities, generally offered free of charge by the cable system. Also see *access channels.*

local loop service: historically, telephone communications path used to transmit telephone and data communications within a community. Term now used also to refer to local distribution of data, whether by telephone circuits, cable TV, or other technlogy.

local origination: term usually used for programming produced by the local cable operator. It may also be film or videotape produced elsewhere and sold or leased to the operator. Term is sometimes used interchangeably with *local programming.* Also see *origination cablecasting* and *access channels.*

local programming: collective term which may encompass not only cable system local origination, but also public, government, and educational access programming. Term is sometimes used interchangeably with *local origination.* Also see *access channels.*

local signals: television signals received at locations within a broadcaster's service area. Term also sometimes used interchangeably with *must-carry signals* which are required of cable systems by the FCC.

low power television (LPTV) station: a type of station with greatly reduced power compared with a conventional station, resulting in a much smaller Grade B contour and broadcast service area; it is licensed by the FCC in a manner similar to a television translator station, except that an LPTV station may originate programming. LPTV stations area not classified by the FCC as must-carry signals for cable systems.

LPTV: *low power television station.*

Major Markets: most often refers to the top 100 television markets in the United States. Sometimes used to refer to the five or ten or so of the very largest television markets in the country. Also see *Top 100 Markets.*

mandatory carriage: television broadcast signals that a cable system must carry in accordance with FCC regulations. See also *signal carriage rules.*

master antenna television (MATV): coaxial cable distribution system of signals received by a master antenna, serving one building or adjacent groups of buildings (such as apartments, hotels, or hospitals). When a satellite dish is used in conjuction with the master antenna, the service is often referred to as satellite master antenna television (SMATV), or mini-cable. MATV and SMATV systems are distinguished from cable television systems by the absence of any local franchise or regulation. Such regulation is avoided by not placing cables over or under public streets or rights of way. Such a system offering a full range of services may be a cable television system for purposes of regulation by the FCC or for purposes of the Copyright Law if more than 50 subscribers are served and the buildings served are not commonly owned, managed, or controlled.

MATV: *master antenna television service.*

maxi-pay service: full service pay programming; from eight to twenty-four hours a day of continuous programming that includes first-run movies, plus, in many cases, sports specials and entertainment features. Compare with *mini-pay service.*

MDS: *multipoint distribution service.*

megahertz (MHz): a unit of frequency equivalent to one million hertz or cycles per second.

microwave link: a relay station in a microwave relay system that receives, amplifies, and retransmits signals. See *microwave relay system.*

microwave relay system: line of sight, point-to-point transmission of television and other signals at high frequencies, by means of geographically spaced microwave links.

microwaves: the radio frequencies above 1000 megahertz. Used in point-to-point communications.

Midwest Video I: refers to the court challenge to the FCC's 1969 action requiring cable systems with 3500 or more subscribers to originate programming. In 1970 Midwest Video Corporation appealed the action. The company obtained a jucidial stay of the mandatory aspects of the rules and successfully prosecuted its appeal before the U.S. Court of Appeals, Eighth Circuit. The FCC took the matter to the Supreme Court and in 1972 a closely divided Court upheld the FCC's authority, but the FCC abandoned the rule two years later. Later, Midwest Video Corporation was to challenge the FCC's cable access and channel capacity rules and usage of the term *Midwest Video* now usually has reference to the more recent case; see *Midwest Video II.*

Copyright 1980, 1983 by Communications Press, Inc.

154 / Glossary

Midwest Video II: refers to the court challenge to the FCC's rules on cable access and channel capacity brought by Midwest Video Corporation. In February 1978, the U.S. Court of Appeals, Eighth Circuit, in *Midwest Video Corporation* vs. *FCC,* set aside these rules as exceeding the FCC's jurisdiction. In April 1979, the Supreme Court, in *FCC* vs. *Midwest Video Corporation,* affirmed the Eighth Circuit's decision. See also *Midwest Video I.*

mini-pay service: pay programming available for a limited number of hours a day; cost of this service to subscribers is about half that of the maxi, or full, service. Compare with *maxi-pay service.*

modem: modulator/demodulator device, such as is used, for example, to connect a home computer to an ordinary home telephone. Also see *modulation.*

modulation: process whereby original information can be translated and transferred from one medium to another, each capable of duplicating the pattern of amplitude and frequency of which the signal consists. Also see *modem.*

monitor: a video display unit having no radio frequency tuning capability. A monitor may be used, for example, to view a TV picture directly from a TV camera.

MSO: *multiple system operator.*

multiple address: transmission from one single point to a number of selected specific points. Compare with *discrete address.* And see also *point-to-point service.*

multiplexing: the combining of two or more signals into a single transmission from which the signals can be individually recovered.

multipoint distribution service (MDS): a common carrier microwave radio service authorized to transmit private television and other communications. MDS provides an omnidirectional signal in the 2150-2162 frequency range, and the service may carry up to two full video channels, depending on the market. The service has proved to be an effective means of delivering pay TV programming especially to apartment buildings and hotels. Pending before the FCC (as of October 1982) is a proposal to reallocate a portion of the 2500-2690 MHz band to make more channels available for MDS, thereby making provision for multi-channel MDS systems.

multiple cable system: a system using two or more cables in parallel to increase information-carrying capacity.

multiple system operator (MSO): an organization that operates more than one cable television system.

must-carry signals: those broadcast signals which the FCC requires a cable system to carry; based on fixed mileage zones, audience surveys, market location, and TV stations' signal contours, in relation to the particular cable TV system's service area.

narrowband: a relative term referring to a system which carries a narrow frequency range (sometimes used to refer to frequency bandwidths below one MHz). In a telephone-television context, telephone would be considered narrowband (3 kHz) and television would be considered broadband (6 MHz).

Glossary / 155

narrowcasting: transmission of information by electromagnetic means, intended for a particularly audience (for example, industrial TV, special-audience cable TV, and business and professional programming). Compare with *broadcasting*.

network: a national, regional, or state organization that distributes programs to broadcasting stations or cable television systems, generally by interconnection facilities. The term has generally referred to the three major television networks, but by 1980 many new networks had emerged—commercial, broadcasting and cablecasting, pay TV, religious programming, etc.—such development greatly facilitated by the availability of satellite transmission.

noise: the accidental, unintended, and normally unwanted components of information received or transmitted as electrical impulses. These interferences degrade the transmission of the desired signal (for example, "snow" in a television picture).

non-broadcast channels: see *cablecasting*.

nonduplication: usually refers to the FCC's network program nonduplication rules, which require cable operators to protect the network programming broadcast by local affiliates by blocking out the programming carried simultaneously by distant stations which the system also carries. This rule still exists, although the term "nonduplication" sometimes also refers to the FCC's syndicated program exclusivity rules which the FCC has eliminated.

optical fiber: thin fiber of very pure glass highly transparent, used for transmitting information by means of light. There is some use of optical fibers in cable television systems today for specific purposes and their use may become more common in the future. Optical fibers have potentially a very high capacity for carrying information, but this is not realized in practice at present.

orbital position: location of a geostationary satellite in orbit at a fixed point in relation to the earth. Used interchangeably with *orbital slot*.

orbital slot: see *orbital position*.

ordinance: law enacted by cities, towns, villages, and other local governmental entities. For cable television, local governmental entities usually first enact "enabling" ordinances, which set forth the terms under which applications will be submitted and a franchisee chosen, as well as general terms of service. The second step is the "granting" ordinance, usually called a franchise, which grants authority to an applicant to operate the system and sets forth specific conditions.

origination cablecasting: as defined by the FCC, programming, exclusive of broadcast signals, carried on a cable TV system over one or more channels and subject to the exclusive control of the cable operator. This is the programming on cable to which the FCC applies the fairness doctrine and "equal time" rules.

pay cable: pay television programs distributed on a cable television system and paid for at additional charge above the monthly cable subscription fee. Fee may be levied on several bases: per-program, full service, tiered service, etc. See also *pay television*, *maxi-pay service*, and *mini-pay service*.

Copyright 1980, 1983 by Communications Press, Inc.

156 / Glossary

pay television: a system of distributing television programming either over the air, by MDS, or by cable, for which the subscriber pays a fee. The signals for such programming are scrambled to keep non-paying persons from receiving service, and a decoder is used to allow the paying subscribers to receive the pay television programming. Sometimes called Premium TV. See also *pay cable* and *Subscription TV*.

penetration: in areas where cable TV is available, the percentage of households subscribing to the service. Also known as saturation.

pirating: 1)making copies of copyrighted material for sale without a license from the copyright holder to do so; 2)receiving for any purpose pay-telecommunications services without making payment.

plant: the physical equipment, buildings, etc., of a broadcast station or cable system.

point-to-point service: the transmission of a signal (audio, video, or data information) via a technology such as microwave, cable, or satellite, directly to the desired receiver(s) rather than to the general public. The service can be in the form of either discrete address, from one single point to another single point, or multiple address, from one single point to a number of selected specific points.

pole attachment rights: the rights obtained by cable TV systems to attach cables to poles owned by telephone or power companies.

portapak: a relatively inexpensive, portable, battery-operated videotape recorder and camera ensemble.

premium radio: pay-radio service, which in 1982 was just in the beginning stages of operation.

premium TV: see *pay television*.

prime time: generally, the broadcast period(s) viewed by the most people, and for which a broadcast station charges the most for advertising time. As defined by the FCC, the five-hour period from 6 to 11 p.m., local time, except that in the Central Time Zone the relevant period shall be between the hours of 5 and 10 p.m., and in the Mountain Time Zone each station shall elect whether the period shall be 6 to 11 p.m. or 5 to 10 p.m.

production: the preparation and recording of a program for broadcast or cable transmission.

public access channel: a channel on a cable system dedicated for use by the public on a non-discriminatory basis, usually with no charge for channel time. Also see *access channels*.

raster: the scanned (illuminated) area of a television picture tube.

receiver: electronic device which can convert electromagnectic waves into either visual or aural signals, or both. For cable TV, usually the subscriber's television set.

redundant cable: the unused cable(s) in a multiple cable system, capped off and reserved for future use as greater channel capacity is needed or as a back up should the need occur.

regional channel: a cable TV channel dedicated for regional programming, usually involving interconnected cable systems.

Registration Statement: in cable television a filing with the FCC, which provides authority to carry television signals if such carriage is consistent with FCC rules. Replaces Certificate of Compliance.

retrofitting: the adding of additional equipment to or rebuilding sections of a cable distribution system after it has been installed. Sometimes needed to increase channel capacity or to be able to provide interactive service.

satellite: orbiting space station primarily used to relay signals from one point on the earth's surface to another.

satellite master antenna television (SMATV): see *master antenna television system*.

saturation: in areas where cable TV is available, the percentage of households subscribing to the service. Also known as *penetration*.

scanning: process of breaking down an image into a series of elements or groups of elements representing light values and transmitting this information in time sequence.

scanning line: a single continuous narrow strip of the picture area of a television tube containing highlights, shadows, and halftones, determined by the process of scanning. Also see *field*.

scramble: to break up an electronic signal into its various component parts. In pay television, for example, the signal is scrambled, and a decoder is necessary for the signal to be unscrambled so that only subscribers would receive the proper signal. A given scrambling encoder of course requires a compatible unscrambling decoder for the system to work.

side band: a band of frequencies adjacent to the carrier frequency, which is generated by modulation of the carrier, and which carries the desired information to be transmitted.

signal-to-noise ratio: the ratio of power level of the desired signal (especially television signal) to undesired noise power level present in the signal.

signal: (1) the message to be transmitted; (2) the electric impulse derived from and converted to the message being transmitted, whether audio, video, text, remote control, or other information.

signal carriage rules: the FCC rules covering the carriage of television broadcast signals on cable television systems. The FCC has no signal carriage rules for cable carriage of radio stations. Also, the FCC preempts states and cities from regulating cable carriage of broadcast signals.

slow scan television: television transmission for transmitting still pictures at a slow rate, using telephone lines or other channels having limited information-carrying capacity. Some uses are for surveillance and for graphics transmission.

SMATV: *satellite master antenna television.* See *master antenna television system*.

158 / Glossary

software: the working materials from which a program is created that will be played out, such as script, audio or visual aids, etc., especially created for the program, and knowledge of how to use the equipment to produce the program. Compare with *hardware*.

specialty station: as defined by the FCC, a commercial television broadcast station that generally carries foreign-language, religious, and/or automated programming in one-third of the hours of and average broadcast week and one-third of weekly prime time hours.

star-configuration: design for a telecommunications system in which signals from the central source are generally transmitted directly to each subscriber (as distinguished from a tree-configuration with trunk and branches to reach the subscriber). The telephone network is an example of a star-configuration.

STV: *subscription TV.*

sub band: portion of the frequency band (5 MHz to 50 MHz) often used for upstream transmission on a two-way cable system. Most subscriber-service networks use the spectrum between 5 MHz and 30 MHz for upstream transmission, with the spectrum between 30 MHz and 50 MHz, called "cross-over," unused because filters with perfect bypass characteristics cannot be built. (The spectrum from 50 MHz to the upper transmission limit would be designated for downstream transmission.)

subcarrier: a carrier which is itself imposed upon another carrier. Independent television, telephone, or other signals can be carried on different subcarriers imposed upon the same carrier.

Subscription TV (STV): form of pay service delivered over the air by scrambling a television broadcast signal.

superstations: independent television broadcast stations whose signals are available to cable systems throughout the country by satellite.

special service network: see *institutional network*.

switched system: a communications system (such as a telephone system) in which arbitrary pairs or sets of terminals can be connected together by means of switched communications lines.

synchronous satellite: see *geostationary satellite*.

syndicated program exclusivity rules: the FCC rules, now eliminated, which required cable operators to black out programs carried on distant television broadcast stations, under certain circumstances, to protect program syndicators' contractual exclusivity with local broadcasters.

syndicated programming: according to the FCC, any program sold, licensed, distributed, or offered to television station licensees in more than one market within the United States for non-interconnected (i.e., non-network television broadcast exhibition. More broadly speaking, any television programming not distributed by one of the major television networks.

teleconferencing: real-time "meeting" of individuals or groups in two or more locations, via video and/or audio hookups, Usually signals are transmitted via satellite and/or telephone lines; some teleconferencing takes place on a few interactive cable TV systems. (The term is sometimes used to encompass interactive telecommunications through computer as well.)

teletext: see *videotex*.

television channel: the range, or *band*, of the radio frequency spectrum assigned to a TV station; the U.S. standard bandwidth is 6 megahertz.

television translator station: a low-powered FM or television station receiving broadcast signals and retransmitting them on a new frequency. Authorized by the FCC especially for difficult geographical locations.

terminal: (1) generally, connection point of equipment, power, or signal; (2) usually referred to in cable TV as the point of connection between a cable drop and a subscriber's receiver, although the term is also used for any "terminating" piece of equipment such as a computer terminal.

tiered service: term used to refer to different packages of programs and services on cable TV systems for different prices; a marketing approach that divides services into more levels than simply "basic" service and one or two "pay" services. Also see *pay cable* and *basic cable service*.

time-division multiplex: technique by which two signals are transmitted on the same transmission medium; each signal is assigned to a separate and distinct time slot within the medium. Also see *frequency-division multiplex*.

Top 100 Markets (Major Markets): for regulatory purposes, the one hundred largest television markets as defined by the FCC. For advertising and other purposes, the one hundred largest television markets as defined by the ratings services.

transmission: the sending of information (signals) from one point to another.

transponder: one of several units, or "addresses," on a communications satellite capable of receiving and transmitting a full video signal. Some users lease the right from the satellite operator to use an entire transponder; others lease only a part of a transponder's capacity (such as for data communications).

translator: see *television translator station*.

trap: a device for removing a set of frequencies from a specified band of frequencies. A *positive trap* removes an interfering signal that has been intentionally introduced into the signal in order to scramble it to prohibit unauthorized reception. A *negative trap* removes a television signal itself, to prohibit unauthorized reception.

tree-network: a design for a cable system in which signals are disseminated from a central source. The configuration resembles that of a tree, in which the product from the root (headend) is carried through the trunk and then through the branches (feeders) to the individual stems (drops) which feed each individual leaf (terminal). Also see *hub-network*.

trunk: the main distribution line leading from the headend of the cable

160 / Glossary

television system to the various areas where feeder lines are attached to distribute signals to subscribers.

turn-key: usually refers to installation where for a fee a contractor builds everything necessary for a complete system.

TVRO: Television Receive-Only Earth Station. See *earth station*.

twisted pair: the wiring used by the telephone system, with a capacity roughly limited to voice and low- and medium-speed data.

two-way cable system: a cable system capable of carrying information both downstream from the headend to any subscriber's terminal and from any subscriber back to the headend. The information transmitted could be of varied forms, such as audio, digital, video, or combinations thereof. Depending on the construction and sophistication required in a given cable system, the system can operate on either one or multiple lines. Also known as *interactive cable*.

UHF channels: the ultra high frequency part of the spectrum allocated for television broadcasting, comprising channels 14 through 83.

uplink: ground-to-satellite transmission. Compare with *downlink*.

upstream: in a cable system the direction from the subscriber terminals to the headend. Compare with *downstream*.

VCR: *videocassette recorder*.

vertical blanking interval: the unused lines in each frame of a television signal (which can be seen as a thick band when the TV picture rolls over) usually at the beginning of each field, which instruct the TV receiver for reception of the picture. Some of the lines can be used for teletext and captioning.

VHF channels: the very high frequency part of the spectrum allocated for television broadcasting, comprising channels 2 through 13. (The VHF band also includes the entire FM band.)

videocassette: videotape in container that provides automatic threading (tape moves in a continuous loop, rather than reel-to-reel).

videocassette recorder (VCR): electromechanical device used to record television sound and picture on magnetic-coated tape in a container that provides automatic threading, for playback on a television receiver or monitor.

videodisc: phonographic record-like device used for playing back prerecorded video (with sound) programming.

videotape: plastic tape with magnetic coating, used to record (and rerecord) and playback video and audio signals.

videotape recorder (VTR): electromechanical device used to record television sound and picture on magnetic-coated tape for playback on a television receiver or monitor.

videotex: the generic term used to refer to a system(s) for the delivery of computer-generated data into the home, usually using the television set as the display device. Some of the more often used specific terms are "viewdata" for telephone-based systems (narrowband interactive systems); "broadcast teletext" for broadcast systems (with digital information stored in the vertical blanking interval, that is, the unused portion, of the television signal); and "wideband broadcast" or "cabletext" systems (utilizing a full video channel for information transmission); and "wideband two-way teletext" (which could be implemented over two-way cable TV systems). In addition to the systems mentioned here, hybrids and other transmission technologies (such as satellite) could be used for delivery of videotex services on a national scale. See also *home information utility*.

VTR: *videotape recorder.*

wideband: see *broadband.*

Copyright 1980, 1983 by Communications Press, Inc.

Related books from Communications Press, Inc.

The Cable/Broadband Communications Book series, edited by Mary Louise Hollowell.
Volume 1, 1977-1978 (ISBN 0-89461-027-9), fall 1977.
Volume 2, 1980-1981 (ISBN 0-89461-031-7), fall 1980.
Volume 3, 1982-1983 (ISBN 0-89461-035-X), fall 1982.
—and the series' predecessor, *Cable Handbook 1975-1976*, also edited by Mary Louise Hollowell (ISBN 0-89461-000-7), spring 1975.

Creating Original Programming for Cable TV, edited by Wm. Drew Shaffer and Richard Wheelwright, for National Federation of Local Cable Programmers (ISBN 0-89461-036-8), fall 1982.

Cable TV Renewals and Refranchising, edited by Jean Rice (ISBN 0-89461-037-6), fall 1982.

International Telecommunications and Information Policy, edited by Christopher H. Sterling (ISBN 0-89461-040-6), spring 1984.

Communications Technologies in Higher Education—22 Profiles, edited by Ruth Weinstock (ISBN 0-89461-025-2, hardcover; ISBN 0-89461-026-0, paperback), summer 1977.

Paperback editions, except where otherwise indicated. This list is current as of December 1984. Inquiries about these and our other communications titles should be directed to Communications Press, Inc., 1346 Connecticut Avenue, N.W., Washington, D.C. 20036.